Leading Your Research Team in Science

Are you full of ideas for new research projects? Could you inspire a research group to achieve great results? This practical guide for team leaders, and those who aspire to become team leaders, offers a unique approach to help develop your research and become a more independent and productive investigator. Learn how to recruit and develop talented team members, how to negotiate contracts and manage your projects, and how to create wider visibility and publicity for your science. From human resources and project finances, legal affairs, and knowledge transfer to public engagement and media performance, this book provides guidance to enhance your skills and combine them with those of support staff on your road to success. With numerous valuable tips, real-life stories, and practical exercises, this must-read guide provides everything you need to take responsibility for leading your own research team. This title is available as Open Access via Cambridge Core.

Ritsert C. Jansen is Professor of Bioinformatics and Dean of Talent Development at the University of Groningen, The Netherlands. He aims to help his early-career colleagues by sharing his own experiences as a researcher. He is the author of *Developing a Talent for Science*, about the life of modern scientists in today's academic environment. He subsequently published *Funding Your Career in Science*, about how to attain research grants. This third book, *Leading Your Research Team in Science*, completes the trilogy. If you want some tips and tricks on how to be an entrepreneur in research, then these books are for you!

Leading

Your Research Team in Science

Ritsert C. Jansen
University of Groningen, The Netherlands

CAMBRIDGE
UNIVERSITY PRESS

CAMBRIDGE
UNIVERSITY PRESS

University Printing House, Cambridge CB2 8BS, United Kingdom

One Liberty Plaza, 20th Floor, New York, NY 10006, USA

477 Williamstown Road, Port Melbourne, VIC 3207, Australia

314–321, 3rd Floor, Plot 3, Splendor Forum, Jasola District Centre, New Delhi – 110025, India

79 Anson Road, #06-04/06, Singapore 079906

Cambridge University Press is part of the University of Cambridge.

It furthers the University's mission by disseminating knowledge in the pursuit of education, learning, and research at the highest international levels of excellence.

www.cambridge.org
Information on this title: www.cambridge.org/9781108701860
DOI: 10.1017/9781108601993

© Ritsert C. Jansen 2019

This work is in copyright. It is subject to statutory exceptions and to the provisions of relevant licensing agreements; with the exception of the Creative Commons version the link for which is provided below, no reproduction of any part of this work may take place without the written permission of Cambridge University Press.

An online version of this work is published at http://dx.doi.org/10.1017/9781108601993 under a Creative Commons Open Access license CC-BY-NC-ND 4.0 which permits re-use, distribution and reproduction in any medium for non-commercial purposes providing appropriate credit to the original work is given. You may not distribute derivative works without permission. To view a copy of this license, visit https://creativecommons.org/licenses/by-nc-nd/4.0

All versions of this work may contain content reproduced under license from third parties.

Permission to reproduce this third-party content must be obtained from these third-parties directly.

When citing this work, please include a reference to the
DOI 10.1017/9781108601993

First published 2019

Printed in the United Kingdom by TJ International Ltd. Padstow Cornwall

A catalogue record for this publication is available from the British Library.

ISBN 978-1-108-70186-0 Paperback

Cambridge University Press has no responsibility for the persistence or accuracy of URLs for external or third-party internet websites referred to in this publication and does not guarantee that any content on such websites is, or will remain, accurate or appropriate.

*In honor and memory of my father,
between the lines he is still present.*

Contents

Introduction 1

1 Team 8

    1.1 Introduction
    1.2 Scout
    1.3 Select
    1.4 Prepare
    1.5 Advance

2 Organization 76

    2.1 Introduction
    2.2 Human Resources
    2.3 Financial Affairs
    2.4 Legal Affairs
    2.5 Patent Affairs

3 Society 136

    3.1 Introduction
    3.2 Open Science
    3.3 Citizen Science
    3.4 Media
    3.5 Web Profile

Epilogue 202
Further Reading 205
Acknowledgments 215
Index 217

Chapter Opener Images

Chapter 1: original image @ iStock/Rawpixel.

Chapter 2: original image @ iStock/asafta.

Chapter 3: original image @ iStock/RapidEye.

Introduction

Who is this book for?

This book is for researchers who lead, or want to lead, their own team to carry out their research plans. It is also for staff who provide support to researchers to help make these plans come true: specialists in human resources, communication, information technologies, finance, knowledge transfer, legal affairs and contracts, learning and teaching, and any other issues important to the day-to-day and longer-term business of research. So what's in it for you? I will outline what you can expect to find in this book here.

- **You're a team leader or aspiring to become one.** You're passionate about science. Your dreams might lie in the exact sciences (biology, chemistry, mathematics, or physics) or in one of the other sciences (medical, behavioral, economic, historical, or social sciences). You're convinced that your ideas will extend the frontiers of your research field. And you dream of having a successful team of dedicated people working with you. How do you make this come true? Today's reality in academia is that it's not enough to be an expert in a particular scientific area; you also need to be multitalented in a range of skills to get your daily academic "business" done. You need to have a nose for the best opportunities and antennas alert for serious threats. You need to be able to "sell" your plans well to attract the necessary funds. You need to recruit the right people and create a team that can work well together. You need to justify to the funding agency and the public why their investment in you is a wise one. You need to know when to protect an intellectual contribution before sharing it openly. There are many more "you need's." Here is a final one: you need to spend your allocated research time as much as possible on doing true research with your team. How can you ever achieve this? *If there is an answer, then it could look like this: research is teamwork, and today's teams are more than ever a union between talented scientific staff and talented support staff. Get to know what these support staff can offer you. Learn to understand their language, and help them to understand yours. Get them truly engaged in your work, and let them support, guide, advise, and educate you and contribute their share to the teamwork. And last but not least, include them in your research successes.* *Take the lead, become entrepreneurial: it is up to you to make your science a successful business.* I trust that this book will trigger you to do just that extra bit, that the passion and performance of your team together will greatly exceed the sum of the individuals' contributions.

- **You provide support services.** Perhaps you work in a university support department. You have great colleagues, in-depth knowledge about processes and procedures in academia, and some basic understanding of what the scientists actually aim for in their research projects. But these scientists are so specialized and sometimes they're rather odd characters too, so it may be a continual challenge to understand and support them. However, it's worth investing time and effort to engage with them and understand a bit about their work. How do you convince them that your expertise can add value to their

3 INTRODUCTION

research business? That together you can make their science affairs work better, hopefully much better? Step into their shoes, figuratively. What is really meaningful to them? What will make them feel well supported? You're an expert not in their field but in your own. Like them, know your environment: what do your peers in support departments at other universities offer? Take a look at the websites of top (and other) universities. Go to national and international meetings on research services, and offer to give oral or poster presentations on your experience. Read the literature – not only business books written in your language but also academic journals such as *Nature, Science, Trends in ...* , *Public Library of Science*, or *Chronicles of Higher Education*. Such journals regularly publish articles about the business of science too. Let your science colleagues know that you are at the forefront of your field, and prove this to them. This aligns you with the scientists. Ultimately, why not consider publishing your department's vision, stories, or best practices in those journals too? It may help to convince the scientists you support and your management that your ideas will make a difference. Now demonstrate your expertise, passion, and ambition to work with your organization's scientists. Researchers need to trust that you do what you say you are going to do (*credibility*), do it when you say you're going to do it (*reliability*), do it in a way that is sensitive for researchers' characters (*intimacy*), do it to serve the interest of the research and researcher and not just your own (*orientation*), and do it well (*effectiveness and efficiency*).[1] *By contributing to the work of successful scientists, their positive recommendations will raise your reputation and justify your role.*

Note: In the rest of this book, "you" refers to the ambitious scientist who wants to make science his or her business by

---

[1] Modified from Maister et al. 2000.

building up an effective team of coworkers. If you are not a scientist but a support staff member, then every time you read "you," it is an invitation to step into the shoes of an ambitious scientist.

How is this book structured?

This book has three chapters to help you do all of this and to do it well:

Chapter 1　**Team** describes how you can build up an effective team of Master's and PhD students, postdoctoral researchers, and other staff, such as technicians. Join forces with staff from human resources to optimize your most important research resource: the people.

Chapter 2　**Organization** describes the formal organization of your team: how to manage the finances, protect intellectual property, and negotiate legal contracts. Learn about the official rules, procedures, and processes of your institution. Join forces with staff in financial, legal, and human resource services, as well as in the knowledge transfer office.

Chapter 3　**Society** describes how to address your peers, the public, and other parties who want or need to know about your project and its results. Enhance the visibility and publicity of your work by allowing open access to information, by involving non-academics in the work, and by using traditional and modern media to share outcomes. Join forces with staff from information technology, library services, and communication services.

FIGURE INTRO.1 The three parts of this book

The three chapters and the various sections can be read sequentially or be dipped into ad hoc. Read and use the information in your own way. The "messages" in the sections are illustrated by *anecdotes* from starting, consolidating, and advanced researchers as well as from alumni who work outside academia: stories can speak louder than anything else. All these stories are presented in the first person, and some details have been changed to protect the privacy of the people who have shared their stories with me – but all the stories are based on true events. The messages in these sections have also been translated into various *"TRY THIS!" exercises* to help you sharpen your thoughts. You can do most of these exercises on your own, but you may benefit from interaction at a research group retreat, in an ad hoc peer discussion session you organize, or in a course for young team leaders.

The trilogy

This book is the third part of a triology for "early career researchers at work." The first book, *Developing a Talent for Science* (2011), discusses essential behaviors and skills. It offers guidelines on how to develop your own talent, how to use other people's talent, and how to develop other people's talent. The second book, *Funding Your Career in Science* (2013), discusses proposing projects and getting them funded.

It offers guidelines on how to get novel research ideas and convert them into successful project proposals. This third book, *Leading Your Research Team in Science* (2019), completes the trilogy and discusses managing funded projects. It offers guidelines on how to deal successfully with all the associated responsibilities.

Disclaimers

This book offers an introduction to a range of issues in the domains of human resources, finance, and legal experts and others such as lawyers and patent specialists.

First disclaimer: The information in this book does not aim to replace financial, legal, or any other professional advice. If you have questions, do contact a specialist, e.g., a human resources officer, project controller, or patent lawyer. How specific issues should be dealt with will differ between countries, between universities in the same country, and sometimes even between different institutes at the same university. Once you have read this book, however, you'll have had a chance to think about such issues and be better equipped to handle such matters appropriately, knowing some of the general actions you need to take to make things happen.

Second disclaimer: This book is not a scientific text. While reading it, try not to accept or reject anything right away; rather, taste, explore and consider, look up suggested further readings, and form your opinions later, as well as your own vision, strategy, and plans for your science business. Do it your way!

1

"It is better to have one person working with you than three people working for you."

Dwight D. Eisenhower (1890–1969)
Thirty-Fourth President of the United States

Team

1.1 Introduction

There is probably nothing new to you about working on a team. As a PhD student or a postdoc you have seen your supervisor lead a team. With luck, it will have been great fun (e.g., taking in cake for the coffee break or going out for drinks to celebrate a recently accepted article), especially if you worked on a team of people with mixed backgrounds (e.g., celebrating Chinese New Year with all its traditions). You'll remember such events forever. The team may have achieved amazing scientific results – more than were promised in the original project plan and more than hoped for because the team spirit inspired everyone to work together to a higher level. At the same time, you may wonder what the keys factors were behind all the fun and success.

From time to time, working on a team can also be quite frustrating, and you may have seen this side of things too (e.g., a PhD student bogged down in details or suffering a burnout or a supervisor who was out of office for too long when you urgently needed his or her input). There is a Jewish and Arabic saying, "Wish your enemy a lot of staff." Disagreements, controversies, and discomfort – you have probably come across some (or maybe too many) such difficulties. Your supervisor may have resolved the issues and created a culture in which problems could be discussed and sorted out. Or not …

Once you start your own team, leading your own people is no longer the responsibility of your boss – but *yours*. The fun is yours, and the hard part is also yours to deal with. Before this point, work was all about making your career. Now it's also about the careers of others. You are welcoming team members and raising them as your "scientific offspring." You are a specialist in your field, and now, more than ever, you need to take care of the human dimensions of your team.

How you deal with the human factor is crucial for making science your successful "business." The following four sections address the different consecutive phases from identifying the first person you recruit to building an effective team:

- **Scout.** Where and how do you attract talented PhD candidates or postdocs to apply for a post on your team?
- **Select.** What criteria should you use when comparing applicants? How do you value and treat applicants equally who can have quite diverse expertise, experience, and characteristics?
- **Prepare.** What good research practices should you and your team members take on board? How do you make them aware of the need for high standards in their research? Can you pay attention to these standards from the very first moment?
- **Advance.** What makes a group of individuals into an effective team? How do you train members well for taking their next steps? How about yourself? Once some members leave your team, how do you restore the team's equilibrium and get back to a state of performing well?

1.2 Scout

Team members, such as PhD candidates, postdoctoral researchers, and maybe a research assistant or two, fly in, stay for a time, and fly out again. They are each present for some of the cycle of starting, running, and completing your projects. This requires you to become a skilled talent scout who can attract and select exactly the right people to fit into your team and work plans. You can start looking for candidates when you have a vacancy, and you may often need to fill the vacancy as quickly as possible in order not to lose the funding. It would be just too bad if you needed to recruit your new member in a rush from a rather limited pool of not quite good enough candidates. Why not start looking for candidates a bit earlier? After all, a good project also takes months to develop from the initial idea to the funded proposal. You can use those months to build up a list of PhD and postdoctoral candidates who may be looking for new positions in due course. As soon as you can start recruiting, you can alert the people on your list and then post job ads and use complementary strategies to extend your list. With such a combined medium- and short-term strategy, the chances of recruiting the "best people in the market" for your growing team are a lot higher. If you start scouting late, time may be ticking away, and you may need to appoint a doubtful candidate in order not to lose your grant money – this would be too bad.

How to attract the best candidates

It is worth running that extra mile to attract and recruit the best candidates for your team. Figure 1.1 shows some strategic options you might use to get in touch with high-potential candidates directly or indirectly. See which strategies work for you.

You can share job advertisements in the traditional way:

- **Job post to your peer network.** Develop, cultivate, and exploit relationships with national and international colleagues and project collaborators to mutually share job openings and opportunities for exchange (internship) programs. Become a member of relevant academic societies in your field, and post your positions on their job portals and in their newsletters. Your university may also be associated with other universities offering additional peer networks to which to post your jobs.

- **Job post at a conference.** Try to agree that you and your colleagues will routinely promote each other's job openings during visits, workshops, conferences, or wherever else you and they go – this is "direct marketing" to well-defined groups.

FIGURE 1.1 Scouting strategies: traditional job ads (top), job ads, and social media (center, left); building up a list of candidates before you post any job publicly (center, right); building up a list of candidates you have met in person (bottom)

A joint advertisement for your positions with those of your colleagues can demonstrate the ambition and dynamics of your group, institute, and university.

- **Job post in a scientific journal.** Publishers such as Nature Publishing Group allow for online or printed job posts. This is an expensive option unless your vacancy is advertised jointly with a number of other vacancies from your university.
- **Job post on an academic job portal.** Your university will advertise your vacancies on its site. National or international unions of universities, academic medical centers, and research institutes may also post their vacancies on a joint web portal. Funding agencies may want to advertise your positions on their websites and in their newsletters.
- **Job post on your personal website.** All serious candidates will look at your website. They will form an opinion about you and your team and whether they would fit in well. Even better, on your website you can show that you have happy group members with diverse backgrounds and share who has worked with you in the past. Be explicit that everyone (different genders, ethnicities, handicaps, etc.) are really welcome (include a broad welcoming statement in all your recruitment methods).

You can, of course, also share job posts by using social networks:

- **Job post on social media.** Tweet your vacancies, and your followers may retweet them. If you inform an influential player in the field, your tweet may spread even wider. Post your vacancies on Facebook, and others may "like" your post so that their friends see it too. Try posting on well-read forums and other social media platforms such as those of your peer community, academic societies to which you belong, PhD and postdoctoral researcher associations, or any funding agency that is sponsoring your project.

- **Web advertisement campaign.** Seek help from support staff to start an online advertisement campaign; for example, with the right combination of keywords, Google ads may be affordable. Your ads may pop up on the screens of good candidates. Other platforms such as LinkedIn can also email or show ads to the right candidates.

- **Headhunting software.** Commercial platforms such as LinkedIn offer free or paid tools for headhunting. Advanced platforms such as ResearchGate and Elsevier's Scival mine publication databases and other online information to find and rank candidates. Candidates are analyzed based on text in their publications, citations, altimetry, and more (e.g., demographics, publication time frames). Again, seek help from support staff.

You can start scouting around before you actually have a vacancy. You can become your own headhunter. Every time you run into someone interesting, consider that person from a recruitment perspective. Note his or her name and some details. At a later stage, you can then email them your job ad. You can ask them to circulate it or send it directly to anybody who may be interested and suitable for the job. Of course, it becomes unethical when you try to acquire someone who works (happily) somewhere else. After all, the supervisor of that person is also your peer. But perhaps they may be looking for a new position themselves.

- **Authors, awardees, and grantees.** Analyze the author list and their contributions to a striking and possibly influential paper in your own field or a neighboring field. A Master's student who coauthored a paper could become your next PhD candidate; a PhD candidate who wrote an article could become your postdoc. If you hear of people who graduate with distinction or who obtain personal grants for internships or awards for their theses, posters, presentations, or articles, add their names to your list. Or perhaps you have served on a grant review or award nomination panel and have seen some excellent candidates. Or you may see a press release about people being elected for membership of prestigious young societies, all more names to add to your list. Be ready when they are looking for new positions.

- **Ambassadors.** Current or former team members also can help to attract several excellent PhD candidates or postdocs from the place where they studied or worked. They can be your best ambassadors, telling others that working with you is great. If some of your alumni now lead their own teams elsewhere, this can generate more candidates for your team when they are looking for new positions. It's worth keeping in touch with your alumni.

- **Career fair.** Delegates of your university may travel to career fairs to interview candidates for you and your peers. These candidates may be in their final year of their current contract and may be starting to look seriously for their next job opportunity. In the weeks before the career fair, you can prescreen CVs from the fair's database and select one or more candidates to be interviewed by your university's delegate during the fair. You can plan further actions for those candidates who pass the interview. For example, you can send them a direct email to inform them that you may have a job opening in a couple of months (or straightaway if the fair is fortuitously held at the time you need to start recruiting).

As a medium- to long-term strategy, you will want to meet some candidates and invest in a personal relationship. They may feel pleasantly surprised and even flattered by your attention and may then seriously consider a job with you rather than with someone else they don't know so well. Be ahead of your competitors by having direct contact with candidates.

- **Student conference.** Help student or postdoc associations organize international conferences at their level, and offer a topical satellite lecture or workshop or a mentoring session at a satellite career fair. You'll get to know the talent pool, and they will get to know you (and hear about future job opportunities at your place).

- **Summer school.** Organize an annual summer school together with some colleagues. Invite some well-known lecturers. Allow Master's students, PhD candidates, and/or postdocs to apply. Possibly you can also offer some support for travel and housing expenses.

- **Research assistant program.** Each year try to offer one or more local Master's students a paid (part-time) job as a research assistant for a couple of months. It's a job, so they won't earn study credit points. Instead, they get a (small) salary that they can use for covering study and living expenses. A job at a bar at night would be much less useful for establishing their career in science. The funding for this might come from your own grant money, or if you're lucky, your institute may have a budget available. Invite local students to apply, and carry out a genuine selection procedure to recruit the best. This can be seen as useful experience to prepare them for future job applications and a reality check for their suitability for a career in academia. It can also give you practice in seeking out the best candidate.

- **International training network.** Coordinate with international colleagues to design a Master's student training program and get it funded, for example, by the EU's Erasmus Program. Master's students from all over the world will apply to prestigious programs. Selected students stay for six months at the place of one coordinator, before they move on to another coordinator and place. As a lecturer and coordinator, you'll be able to assess which students are best qualified for your job vacancies.

- **Short-term stipend.** Offer promising external candidates from your scouting list a stipend for a short stay of a few days or perhaps even two weeks in your group. Some institutes or consortia have budgets available for exchange visits between partners. It's good for the candidate's curriculum vitae (CV), and both of you can test whether the match is mutually beneficial. In addition, your team members can tell you how they like interacting with the candidate. See Box 1.1 for an example of an invitation for a short stay with your group.

BOX 1.1 Invitation for a three-day site visit

> Dear Mr./Ms. [name, e.g., Johnson],
> Thank you for the Skype interview last April 15th for one of my PhD/postdoc vacancies.
> It is my pleasure to invite you, along with a few other candidates, for a campus visit on May 10–12th. Please plan your trip to have these days available for your visit. I would like to invite you to attend the Faculty's Research Seminar (click here for the speakers) on May 10th at 3 P.M. A week beforehand you will receive details about the prep work required for the seminar. Further program details will be made available in due course, but your visit will end on May 12th by 2 P.M.
> After the site visit, we will make a final selection of candidates based on our criteria of academic skills (e.g., potential to work with digital tools and ability to analyze data) and social skills (e.g., ability to work on an international and interdisciplinary team).
> We can offer you a travel stipend (which you can add to your CV) to cover your travel costs (economy class) and accommodations. A hotel room will be booked for you once you have confirmed the dates.
> Thank you for your interest in the position. I would appreciate it if you could confirm these dates as soon as possible, but by April 25th at the latest. We look forward to hearing from you and to meeting you soon!
> Yours sincerely,
> [Your name and degree, e.g., Maria Dunn, PhD]

Choosing between a PhD and a postdoc

You may have built up a nice long list of high potentials, some of whom are still Master's or PhD students, while others may already be postdocs. How great it would be if you could welcome one or more people from your list to your team. However, before you move on from scouting to selection, do note some of the advantages and disadvantages of recruiting a PhD student or a postdoc.

- **Ideal group composition.** Educating PhD candidates is a primary task for any university researcher; you are probably expected to supervise PhD candidates on your team. You have too many ideas to follow up yourself, and one or two PhD candidates can work with you and multiply your research capacity. Because they stay for three or four years with you, it pays to invest plenty of time in training and supervising them. However, you will find that you have limited supervision capacity; if your group is going to grow more, so will your role as supervisor. At some point cosupervision of PhD candidates by a postdoc will be welcome. You will then be able to leave for a longer trip or take a holiday while someone else takes care of the daily supervision and some of the daily group dynamics. There is also a benefit to spreading training over more people because postdocs can help you train junior members and improve the junior/senior staff ratio. Teaching student courses is another activity you can share with a postdoc. The cosupervision of PhD candidates and the gain in teaching experience will strengthen a postdoc's CV, but you should explicitly credit them for taking on additional responsibilities (and successes). However, for postdocs to succeed, they need to demonstrate their scientific independence, and they will therefore need to work increasingly on their own ideas rather than on your projects.

- **Skills comparison.** A PhD student may start as some sort of research trainee in your group but should develop scientific independence over the course of the project. The "should" in this sentence refers to the risk: not all PhD candidates develop their knowledge and skills fast enough, which is particularly relevant for monodisciplinary-trained PhD candidates starting on an interdisciplinary project. Some may need a lot of supervision for a rather longer time than hoped for. But PhD candidates can be flexible if you have the time to train them and help them develop. In contrast, postdocs – if well screened during hiring – are scientifically quite independent from their first day and will generate results more quickly, for example, increase the number of good group papers published, bring critical diversity into your group's discussions, or help in supervising undergraduate and graduate students. But some projects need to start producing results right from the start – there is no time for the two-year growth period of a PhD candidate. The tasks, deliverables, and timeline then call for a postdoc with the right background. In the same project, some other tasks may be more suitable for PhD candidates, so if you have enough funding, you could appoint both.

- **Commitment comparison.** A PhD candidate will stay until the end of the research project in order to finish their thesis – it's their entry ticket to a career, whether in academia or elsewhere, and failing is not an attractive option. In contrast, a postdoc position is usually short term and a stepping stone to a more permanent job in academia (e.g., instructor, lecturer, or assistant professor) or elsewhere (e.g., research group leader in industry). The postdoc needs to search proactively for his or her next job and will ask you for recommendation letters. You must help your postdoc leave, even if this means that his or her contract with you will be terminated earlier than practical for your project work. Another point is that when hiring a postdoc, you also hire their history. Often they need time to finish earlier work. It takes time to get manuscripts on earlier work published, and quite often reviewers of these manuscripts will ask for extra work to be done. This is all time your postdoc cannot spend on your project. Or the postdoc will need to work in the evenings and on weekends so that they have no time to switch off and relax.

A YOUNG TEAM LEADER'S ANECDOTE

One virtual handshake away

I have over 300 followers on Twitter. Many of them I have never met and don't know personally. So I tweet my vacancies and hope interested followers will apply. Sometimes it works, but I thought I could test a possibly more effective strategy. Curious? Then read on. I follow some 100 influential scientists, academic groups and organizations, publishers, and funding agencies. It shows where my interests lie, and I am actually looking for people with similar interests. I decided to use an internet crawler, software that allowed me to collect information from websites. This enabled me to download the list of followers for each Twitter account on my list of 300 people following me. And then, with some further data crunching, I derived a list of people who showed a good overlap in terms of who they followed and who I followed. After further analysis, I had a list of 10 top candidates, and I sent each of them a private message explaining how I had found them and what job opening I had. I was definitely the first recruiter to contact them for their next step. And my search strategy impressed them. I conclude that so far this strategy has worked amazingly well. Now I'll have to see whether I can really hire one of them in the next few months once their current jobs finish.

TRY THIS!

- Interview five to 10 experienced researchers about their scouting strategies: what worked well (or badly) and why? Discuss the options in Figure 1.1.

- Ask your human resources (HR) recruitment officer for advice, and discuss the options in Figure 1.1. Has the university subscribed to recruitment software for searching for suitable candidates and advertising jobs (e.g., LinkedIn ads or Google AdWords)? Does it attend specific career fairs with access to a database of participants?

- Make a list of the five keywords to be used as input for scouting tools:

	Keyword
1	
2	
3	
4	
5	

- Use the keywords to mine social network sites, e.g., ResearchGate or LinkedIn or career fair databases for candidates.

- Make a list of at least five influential people and communities on social media such as Twitter and Facebook. Alert them when you're looking for people, and hope for likes, retweets, and more:

23 SCOUT

	Influencer or community	Social media name
1		
2		
3		
4		
5		

- Make a list of at least five important student or other conferences, society meetings, or academic network meetings, and check whether anyone from your organization is attending and can post your job ads:

	Conference	Who's going
1		
2		
3		
4		
5		

- Make a list of at least three top male and three top female candidates you don't yet know by looking for (1) authors of important scientific articles, (2) awardees of relevant minor and major distinctions, and (3) receivers of relevant minor and major grants:

	Article, grant, award	Name of author, receiver, awardee
1		
2		
3		
4		
5		
6		

- Check whether your organization offers summer schools (or winter schools) and how it can help you organize one. Who would you like to co-organize it with you, who would you like to give a talk, and how would you advertise it?

- Use a fraction of your budget to offer talented Master's students the experience of working in your group as your assistant. Ask students to apply formally.

- On your website, offer a stipend for short stays in your group. You may attract just a few more candidates or evoke an avalanche of applications. See whether good candidates apply. Read the invitation in Box 1.1: what assignments can candidates be asked to complete before their visit, and what activities could be organized for their visit? Which agencies fund peer networks for training PhD candidates or postdocs? Can you join a training network? Or can you team up with some strong partners to propose a training network grant application together?

1.3 Select

Get the right people on board with your team and in the right seats. Making a selection is a very serious business, and the golden rule is: in case of serious doubt about a candidate, don't hire him or her. Otherwise, your group will never become an effective team. The implications of a mismatch are huge, for the candidate in his or her career progression, for the team because the funding and timing may not allow for a restart with another person, and for you and your career because you need to demonstrate "sound leadership in the training and advancement of young researchers." A traditional long interview in person or by Skype can be useful in deciding to turn down a candidate you have never seen or spoken to before, but it is still a dangerous basis for making a precipitous positive decision. Although you may feel that you need to start the project as soon as possible, you may be impressed by the candidate, and so on, but a good one-hour session can still lead to a frustrating "marriage in science" for one or more years.

We all tend to evaluate candidates based on their past academic merits and our prediction of their future prospects (Figure 1.2). You will have to work with this person, probably on a daily basis, so don't forget to evaluate their fit into the team. The hard and soft selection criteria can be specific to your research field, your project, and your current team composition and attuned to the level of the position (PhD or postdoc). In addition, there are several evaluation criteria generic

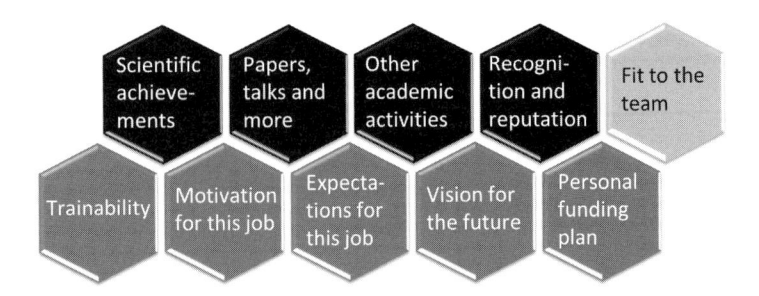

FIGURE 1.2 Evaluate the candidates at various levels: past performance (black), future perspectives (gray), and how well they fit in the team (light gray)

to many fields and projects (again these can be attuned to the level of the position):

- **Scientific achievements.** Your candidates have experienced what it is to do research, as part of their thesis project or a postdoc job. Ask them what they are most proud of. Can they outline in 100 words what their two to three main intellectual contributions were (versus the contributions of the supervisor or coauthors). Ask them explicitly not to focus on what they have done but on what they have created, discovered, invented, developed, or achieved. Also ask them about their failures, disappointments, and frustrations and how they have dealt with these. *Are they independent, creative, and persevering?*

- **Papers, talks, and more.** It's great to be independent and creative, but not enough. Only if candidates present and write well can their peers and society benefit from their work. Ask candidates to show their thesis reports and whether these resulted or will result in a paper. What else have they written, submitted, and/or published; how much of the writing did they do themselves; and how independent were they? How do they perceive the writing process, including making revisions based on reviewers' comments? Let them write a brief piece on the spot as a reality check. Have they ever had experience reviewing a manuscript for a journal editor (perhaps one passed on by their supervisor)? Have you seen or heard them present a poster or paper at a meeting? How convincing and stimulating were they? How lively was their interaction with the audience? Let them give a presentation to your group, and do let them talk with the group and with individuals about their work and your group's work. The quality of the work is what really matters, not the number of papers. *Are they eager and able to communicate effectively at an academic level?*

- **Other academic activities.** Have they supervised students or taught any classes? Did they ever help organize or coordinate anything like a journal club, student conference, satellite meeting, or a conference, or have they acted as an editor of a student journal? Have they been a member of a student council or a student member on a faculty board? Have they worked only locally, or nationally or even internationally? What can they say about their academic network? Who do they know, where have they been, and why? Try to find out whether they have hidden talents that may be of use to your group or that, with some extra investment, could position them well for their next career step. For example, do they like sharing results with the public via press releases, media posts and blogs, interviews, or generating newspaper coverage, e.g., by sending a letter to the editor of a local or national newspaper? *Are they open to helping academia run well?*

- **Recognition and reputation.** Did they get excellent grades for their student exams and, in particular, for their thesis? Was their thesis well received? Did it win an award? Have they ever been given a poster award or a best presentation award? Or a travel stipend to attend a conference, or a personal grant or fellowship? Is their work being picked up quickly by the research community or societal groups, i.e., downloaded, tweeted, highlighted, cited? Have they been invited to give an oral presentation at a workshop or conference? Can they provide you with the names of supervisors and recommendation letters from them? Can you phone the supervisors and discuss their recognition of the candidate's scientific abilities. Is this a high-potential candidate? Would the supervisor hire this candidate with no reservation if in your place? *Does this candidate have a good reputation?*

- **Trainability.** Good past performance doesn't guarantee that candidates will be successful in the future. To successfully contribute to your project, they need to be hungry for new knowledge and ideas, quickly picking up while thinking for themselves, eager to develop the hard and soft skills necessary for the project. A short stay in your group can help you assess their trainability. Can they familiarize themselves with your project's topic, can they connect it to their earlier experiences, and can they perhaps already suggest new ways for your project to progress? Will they contribute to the deliverables and at the same time negotiate to make it their own project that might deliver more than your original expectations? *Is this candidate trainable and taking ownership of self-training?*

- **Motivation for this job.** It can be quite informative to start an interview with an open question such as, "Tell me how you prepared for this meeting?" Has the candidate done his or her homework? Did he or she study your group website, look up the information on your research lines, and maybe read some of your recent publications? Why does the candidate want to do this project? Why with you and/or your team? Is this a well-thought-through career step or the only job ad available? Has the candidate ever turned a job offer down, and why? *Did the candidate prepare well for the application and interview/meeting?*

- **Expectations of the job.** In terms of content, work, career – do the candidate's expectations match yours. Discuss the work plan, potential risks and opportunities, and how they would start on day one. What sort of interaction with the supervisor and group members are they hoping for. Do they need special working conditions (e.g., a quiet room rather than a shared office space or flexible working hours because of childcare, etc.). *Are all the expectations mutually clear, and do they match well?*

- **Vision for the future.** This is very important: what is the candidate truly passionate about? What does he or she see as the most exciting challenges in his or her academic life? What makes the candidate tick as an academic? Does the candidate have clear dreams or even any concrete ideas for a future research line of his or her own inside academia (discuss that universities need only a limited number of professors) or for a future career track outside academia (universities educate the next generation workforce for society)? Can the candidate explain how a PhD or post-doctoral position in your group would serve his or her future career goals? What core or general skills does the candidate want to strengthen? *Does the candidate have a vision of what he or she wants to learn and where he or she wants to go and the (beginning of a) concrete and realistic plan on how to achieve this?*

- **Funding potential.** Does your PhD or postdoc candidate have ambition, ideas, and a wish to bring in some additional funding during the project, or to apply for a grant to cover conference costs, or to fund a small project of his or her own? Any reluctance to discuss such a strategy should make you pause and seriously wonder whether this candidate is the one to hire. You can introduce the candidate into the landscape of available grants (*is this a grant for you?*), help the candidate benchmark himself or herself (*are you ready for this grant?*), and help the candidate make a plan to get ready (*what can you or we do to be more ready in two to three years?*). Perhaps you should consider making the candidate draft a personal funding plan (PFP) as a standard part of the selection procedure. *Is the candidate willing to draft and discuss a personal funding plan?*

- **Fitting into the team.** You'll be working long hours together with your team members. Will you get along well with each other? Have a drink, lunch, or dinner with the candidates to learn more about their personalities: you need to see them when they are relaxed and not stressed or hyped. Let your team members meet the candidates to see how they all get along and to hear how the candidates behave when you're not around. Then discuss with the team their and your views on the fit to the team. People can be quite different, and this can actually be a strong asset: they can complement – or even strengthen – the hard and soft skills and expertise levels already present in the group (more on this later). *Are they a team player and complementing the current team well?*

TABLE 1.1 Personal funding plan

Type of grant	"Is this type of grant for you?"
Travel grant	Visit important conference or a famous scientist to learn about the newest developments.
Personal research grant	Do you have a bright idea for a project?
Collaborative research grant	Join existing collaboration.
Training grant	What do you want to learn and where?
Other types	Go wild!
(National, international, mono- or cross-disciplinary, fundamental or applied, public or private money, anything goes …)	
Benchmarking the candidate	"Are you ready for this grant?"
CV	Almost as strong as other grantees?
Research line	Write a white paper.

31 SELECT

Research impact	Make a video or blog or call a journalist.
Community roles	Organize a workshop.
National/international networks	Contact the important people in the field.
Prizes, memberships	Look for some options, suggest nomination.

(Do your homework, compare yourself to others who have been awarded a grant)

Action plan
"What will you do to be more ready for an application in two to three years' time?"

Action 1	Action 2	Action 3

(Make an *i*SMART action plan: inspiring, specific, measurable, acceptable, realistic, timely)

Note: Items to discuss (left column) and questions you can ask the applicant (right column). See also the author's book, *Funding Your Career in Science* (Cambridge University Press, 2013).

An applicant may write in their application or tell you during the interview that they are creative, independent, and resilient, a good team player who writes fantastic papers in no time. But whether this is true or not, or a little bit true, may be hard to discern from the CV or motivation letter or from a "yes, sure, I'm creative." The five steps of the STARR method (Figure 1.3) can help you get a more objective picture of the applicant (and to select the true star).

- **Situation.** Ask the applicant to describe a situation in which a particular personal or professional skill (e.g., conflict resolution, critical thinking, perseverance, problem solving, time management) was required. Who else was involved, such as one or more group members or external people?

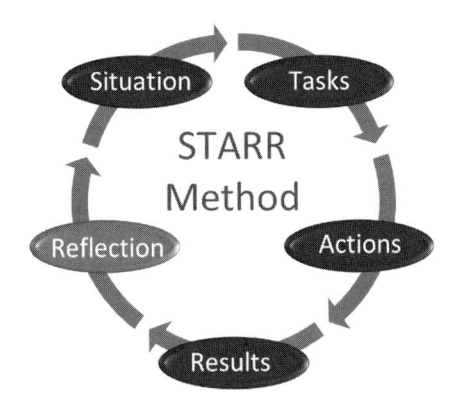

FIGURE 1.3 The five steps in the STARR method

- **Tasks.** Ask the applicant what their task was, how this fitted in the group's tasks, and what the supervisor's and their own success criteria were.
- **Actions.** Here the applicant is asked to describe what they actually did: why, how independently or codependently, and what other skills were needed.
- **Results.** Were the actions successful? The applicant can describe whether the results were as anticipated, fell short, or were beyond all expectations. And how their actions contributed to the team results, hindered them, or helped the team reach a higher goal.
- **Reflection.** Perhaps the most important question in the STARR method is: what went right, what went wrong, what are you uncertain about, what would you do differently the next time you are in a similar situation, and why? Issues will keep recurring until the applicant learns the lesson associated with them. Here the candidates need to show their ability to self-reflect honestly and to learn and improve.

Agree before the interviews with the members of the selection committee on the selection criteria to be used and how to evaluate them (e.g., STARR interview, home work for the candidates, etc.).

Pitfalls during selection

You, as a research leader, want to select from as large a pool of candidates as possible. Unfortunately, there has been a substantial and unfortunate brain drain from academia (Box 1.2).

Selection criteria that are traditionally used in academia need to be revisited. For example, selecting a candidate:

- **For following the standard academic career.** You may look for candidates who want to follow a standard academia career: a Master's degree, followed by a PhD, become a postdoc, etc. However, older people who have had alternative careers or a career break can also be top candidates for PhD and postdoc positions. Some may bring in their own funding, a thesis plan, or even a concept thesis spinning off their current work experience. Outsiders with their "outside" expertise and experience could give a major boost to academia and to your research group in particular. *What makes this senior candidate tick for your group's research?*

- **For having published the most papers.** Some people can and want to work 60 to 70 hours per week; others can't or don't want to. Those who work more hours per week may well publish more papers – but they aren't necessarily the smartest or most creative researchers. For example, one candidate worked on a conservative project for four years and published eight mediocre articles. Another candidate worked on a risky project for four years and published two striking articles. Who should you select? *Ask for the two to three best recent papers (or other output), and evaluate these for (potential) scientific and societal impact.*

BOX 1.2 We are losing talented women and men

> In many countries, student populations have equal numbers of young men and women, whereas the proportions of PhD candidates are slightly less balanced (e.g., 56 percent males versus 44 percent females), and higher up the academic career ladder there is a clear skewing toward men (e.g., in 2018 in the Netherlands, full professors are 80 percent men versus 20 percent women; for university board members, 72 percent are men versus only 28 percent women). If universities recruited the 100 most talented professors, 50 should be men and 50 women, not 80 versus 20. So 30 men became professors, whereas 30 *more* talented women should have been recruited. Worse, perhaps some of the top 50 women decided not to work in academia, and some of the top 50 men too, not because they have no passion for doing research but because they felt they wouldn't fit into the current culture and career system maintained at universities today.

- **For having visited the most conferences.** Parents who take up maternity or paternity leave and have family responsibilities are necessarily less active at the national level and particularly internationally; for example, they may attend fewer conferences and decline invitations to speak more often. This also applies to people with certain disabilities. They and many other people can participate when conference organizers embrace today's digital network technologies. This also saves traveling time, funding spent on hotels and (air) travel, and the environment. *Discuss during the interview finding (or being helped to find) help to attend important meetings and present their work and help the academic world to embrace the virtual conference technology.*

Get the selection panel for a new member thinking along the same lines. Together you should check all the selection criteria for potential bias and adjust them to exclude, or at least reduce, bias. Then agree to consistently apply these criteria throughout the selection process: from advertisement to final negotiations with the top candidate.

Unfortunately, it's not enough to review and adapt the selection criteria: equal opportunity criteria are great but remain ineffective if certain groups of talented candidates do not apply. For example, women tend to be less tempted to apply for high-level positions than men, for many diverse reasons (e.g., cultural, family). Therefore, you need to actively and deliberately encourage women to apply (check your job advert for gender bias: see TRY THIS). And what holds for gender may hold equally (or sometimes more) for people from different ethnicities and cultures, for those who are disabled, and for minorities in general.

It's also not enough to improve on the recruitment procedure: candidates may look beyond the nice words and promises and check whether you practice what you preach by talking with your current or former group members. For example, in general, you'll likely miss quite a few bright people unless your group's culture is warm and helpful toward young researchers who want to start a family, toward team members needing to care for a disabled child or elderly parents, or toward people who wish to combine their work with other activities outside work.

You may unconsciously select and hire new people who are similar to you: this is a well-recognized "more of me" selection mechanism. But you should consider your own personality and what type of candidate would best complement the group members you already have. The "Big Five personality traits" describe the most important factors:

- Extravert/introvert
- Collaborative/competitive
- Organized/easy-going
- Sensitive/confident
- Curious/cautious

If you are an extravert, you may believe that introverted people lack initiative, are socially less competent, overly modest and cautious, and indecisive. Similarly, if you are an introvert, you may believe that extraverted people are rather offensive, pushy, and rude. But introverted people may actually prefer to first tune into the needs of their discussion partner. Extraverted people may actually think and speak at the same time and hope that their discussion partner will do the same. In academia, extraverted personalities seem to have become the norm, which can falsely disqualify introverted but highly talented people. Similar biases exist for the other five main personality traits.

Without diversity in personality and more, we would all think and behave alike. People who clearly differ from you may not have an equal opportunity of being recruited, while their different values, fresh ideas, and alternative approaches may shed completely new light on your research questions and methods (see the following anecdote).

"When science is inclusive, everyone wins."[1]

Bring together and truly include members with different abilities and knowledge skills that are relevant to your project (Figure 1.4).

---

[1] Lee 2014.

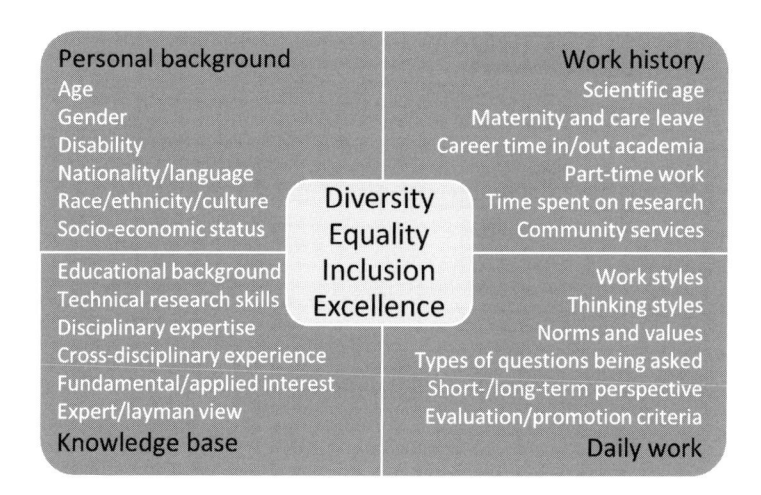

Personal background	Work history
Age	Scientific age
Gender	Maternity and care leave
Disability	Career time in/out academia
Nationality/language	Part-time work
Race/ethnicity/culture	Time spent on research
Socio-economic status	Community services
Educational background	Work styles
Technical research skills	Thinking styles
Disciplinary expertise	Norms and values
Cross-disciplinary experience	Types of questions being asked
Fundamental/applied interest	Short-/long-term perspective
Expert/layman view	Evaluation/promotion criteria
Knowledge base	**Daily work**

Center: Diversity Equality Inclusion Excellence

FIGURE 1.4 Diversity, equality, and inclusion underpin excellence

A GENDER EQUALITY CHAMPION'S ANECDOTE

Shame on me

A professor emailed me and asked, "Can I do a short sabbatical in your group?" Initially I wasn't very keen. He was from a country that I considered to be underdeveloped; I feared he would simply highjack my ideas and contribute very little. I didn't see our research themes as matching particularly well. Okay, maybe, but I felt his request was rather farfetched. My curiosity won, and he stayed with my group for three months. We talked a lot and exchanged ideas; it turned out to be a surprisingly rewarding and pleasant time. He offered me totally different insights and visions on "my" research questions. And the joy was mutual. Together we wrote a perspectives paper on where the field should go, and it was quickly accepted for publication in a highly visible journal. Shame on me, as a woman who is always alert for gender discrimination from male colleagues, not to have noticed how my own prejudice had biased and fooled me.

TRY THIS!

Prepare for the selection

- Make a concrete list of criteria for evaluating the applicants. Rank the criteria for importance.
- Prepare STARR questions for the most important criteria (see Figure 1.3).
- Agree with the panel on the final list of criteria and method for evaluation.
- Decide on the roles during the selection process: who is the chair, who is making notes, who is watching over diversity, equality, and inclusion?

Gender

Some behaviors (and words) are stereotypically masculine, whereas others are feminine.

- Label the features in the following table as more masculine, more feminine, or indifferent. Explain your choices.

Short-term successes	Collaborative	Career promotion
Sharing knowledge	Listening	Winning
Standing in the spotlight	Willing to take risks	Background support
Sustainable relationships	Visionary	Process-oriented
Hierarchical status	Empathy	Task-oriented

- Check your own job advertisements for the number of words that are stereotypically more male or female. Discuss whether this would put women off applying or encourage men to apply; alternatively, is your advertisement okay?
- Run a "gender decode for job ads" tool on your advertisement text (e.g., http://gender-decoder.katmatfield.com/about#masculine).

Daily work

Your top talent candidate happens to have special needs, but could you:

- Arrange an adapted workspace for a person with a physical disability who uses a wheelchair?
- Set up a quiet work room for a person who is sensitive to noise?
- Adapt computer equipment for the work desk and in the lecture room for a person with a visual impairment?
- Arrange a room for a mother needing to express breast milk?

Tests for implicit bias

- Complete one of the implicit bias tests (e.g., related to age, gender, science, career, etc.) at https://implicit.harvard.edu/implicit/selectatouchtest.html. Challenge your selection committee and your current team members to do the test as well and discuss the experience together.

1.4 Prepare

Primer is an undercoat for paintwork to ensure that the new paint adheres well to the old surface. Your new group member also needs preparation, like an undercoat, to become an effective team member. On the first day you will introduce the new member to the group and attend to the practical details of where to sit, how to log into the computer network, who from administration can help, and so on, all toward a smooth transition into their new job. Preparing the new member for their research tasks is as important, if not more so. From the first day, you have to create an awareness of what is often referred to as the "code of conduct for good research practices." Without this undercoat (code of conduct), the paintwork (results from the research) may look nice for some time but will not be enduring. To make your research projects and your team really successful, it is essential to help your team members study *themselves* in the first place and for you to perform self-reflection too. *Refine the most important resource in your research – the people.*

Humans create science, but they can err, be ignorant or inexperienced, short-sighted, or hesitant or reactive; take risks easily or be conservative; follow the mainstream or prefer new trails; and have false beliefs or limiting prejudices. Humans in science are also subject to fierce job competition, up-or-out promotion stress, peer pressure for short-term results that can be published in *Nature* or *Science* with high visibility, or following hypes with a high risk of their results being scooped. Increasing personal interests can make humans opinionated and biased, dominant and arrogant; they can develop strong egos and hold hidden agendas and become selfish and stubborn or even narcissistic and manipulative. Perfectionists and thinkers, achievers, and supervisors who have unhealthy ambitions or who demonstrate unhealthy behaviors can be very harmful to science (Table 1.2).

Make your team members aware of the many pitfalls in the bumpy road to excellence, and let them stay far away from the illusion of excellence (Figure 1.5, left). Surprisingly, it was found that 60 percent of studies published in fields such as medicine and psychology cannot be replicated.[2] An incredibly high number of "landmark papers" published in high-impact journals contributed to this "illusion of knowledge" and

TABLE 1.2 Examples of four personality characteristics and how they may change from healthy to unhealthy behavior

	Perfectionist	**Thinker**
Healthy state	Objective	Visionary
Transition state	Rather rigid	Antagonist
Unhealthy state	Stuck in details	Isolated
	Achiever	**Supervisor**
Healthy state	Goal-oriented	Serving the team
Transition state	Prestige-oriented	Self-contented, territorial
Unhealthy state	Narcissist, making up fake stories	Self-overestimating, dominant

Source: Modified from Riso and Hudson 2017.

---

[2] Special issue, *Nature* (2015); see "Further Reading" for this and other articles, editorials, blogs, and guidelines.

"illusion of excellence." What if 60 percent of all the published studies in your field cannot be replicated? If you read uncritically, trust the literature, and base your research on this body of knowledge, your results are *more likely to be false than true*. If you are uncritical, you may also overlook the *known* or *unknown hidden facts or factors* that do not fit your "story" and would have put your findings in a totally different light, even if they had been replicated (see also the anecdote at the end of this section).

As a team leader, it's your duty and in your own interests to create an environment where your team members aim for real excellence (Figure 1.5, right). You need to prime your new team members and critically monitor yourself and your team as the project evolves. The last part of this section proposes some options of how you could put this theme firmly on your team's agenda.

Being independent and critical

For junior people, it may not be easy to become an independent and critical researcher. These people have taken classes for years, studied many textbooks, completed

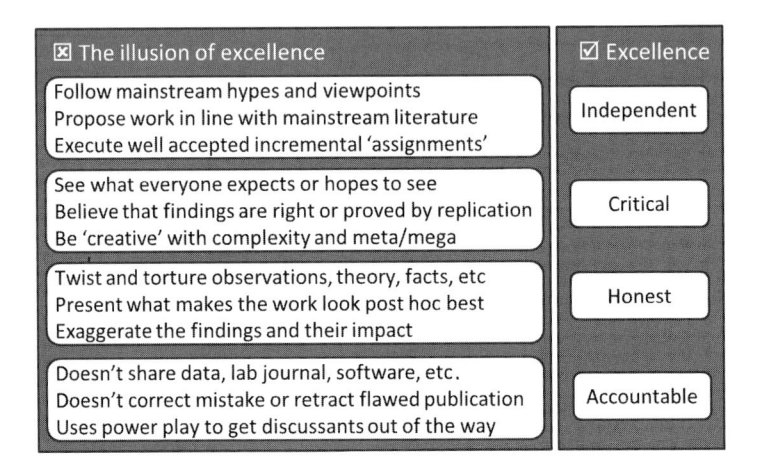

☒ The illusion of excellence	☑ Excellence
Follow mainstream hypes and viewpoints Propose work in line with mainstream literature Execute well accepted incremental 'assignments'	Independent
See what everyone expects or hopes to see Believe that findings are right or proved by replication Be 'creative' with complexity and meta/mega	Critical
Twist and torture observations, theory, facts, etc Present what makes the work look post hoc best Exaggerate the findings and their impact	Honest
Doesn't share data, lab journal, software, etc. Doesn't correct mistake or retract flawed publication Uses power play to get discussants out of the way	Accountable

FIGURE 1.5 Excellence. Science needs excellence (right), not the illusions of successful scientists (left).

hundreds, if not thousands, of carefully set assignments, have passed exams, and have done thesis projects with experienced researchers who gave them well-defined tasks to complete. Why should they doubt what they were taught, have read in textbooks, or were instructed to do by all these knowledgeable lecturers and researchers? Yet this is just what they do need to do – the opposite of what they may have been trained to do so far and the opposite of what is outlined in the left column of Figure 1.5. Critical and independent thinking is, first, the ability to change your own way of thinking. Three simple examples can help open the discussion with your team members:

- **Connect the dots in four lines without lifting the pen from the paper** (Figure 1.6a). It's a well-known puzzle, and some people will already know the solution, whereas others will need a few minutes to solve it. Several intriguing things happen: those who know the solution often turn their attention to something else, such as chatting with a neighbor or checking their cell phone. *Others complete the assignment and then stop thinking too.* This is common practice in education: do what you're requested to do, and you'll be ready for the exam. Typically, a few individuals go beyond the assignment. Some explore opportunities by folding the paper or using equipment such as scissors and copiers. Incidentally, someone may prompt the next question: can we connect the dots with three, two, or even only one line? They take the lead (rather than being led by an instructor and the direct assignment) and go beyond your expectations (rather than being satisfied with the initial solution). *It is this ability to independently, freely, and creatively "further explore the universe" that will break new ground in research.* They don't think inside the box, nor do they think outside the box: for them, there *is* no box. Help your team members to understand that research work is no longer a matter of completing assignments.

- **Peel a banana** (Figure 1.6b). Many people open a banana by the stem. They have always done it this way and never questioned it. A few will open it at the other end and argue that it's generally simpler to open it here than by the stem. It's the approach used by chimpanzees and other apes, they say. This dichotomy in the audience is striking. *Most follow the mainstream approach without ever questioning it.* And some who have discovered that their approach wasn't optimal explain that they were too strongly preconditioned by the past: *reprogramming habits and other behavior is not straightforward*, not even if you want it to be. Whether to open a banana by the stem may be a question relevant to Westerners, who eat their bananas when they are still unripe. People from other cultures may eat really ripe bananas, and then it is simple to open the banana anyway. The question of where to open the banana is totally irrelevant to them; the research question we ask may also be irrelevant if we change values, norms, or habits. Help your team members to understand that they shouldn't take other people's research questions and approaches for granted.

- **Count windows** (Figure 1.6c). Students, PhD candidates, and postdocs typically count between five and nine windows when shown this photograph. Sometimes a smart person sees that each window consists of four smaller panes, so the answer is somewhere between 20 and 36. But generally there is unanimous consensus that the number is large, although there can be debate about counting the panes or subwindows separately. We count and measure a lot in our research, and it can be a surprise that even in this simple case the answers can be so different. But no one takes into account that the windows could be fake. And they are fake for good reasons! This house was built in France, where house owners had to pay taxes proportional to the number of doors and windows in their house (1797–1926 (contribution sur les portes et fenetres). Painted windows, looking like genuine windows, could make you look richer but not

45 PREPARE

FIGURE 1.6 Three simple assignments can help you open a discussion on excellence – or the illusion of it – with your team members; see text for further explanation. (a) Can you connect all the dots with just four straight lines without lifting your pen off the paper? (b) How do you peel a banana? (c) How many windows do you count? Photograph A.K.M. Disler.

cost extra tax. This "you see what you expect to see" is the mechanism adopted by the house owner, and it's very likely to occur on a large scale in science too. Large groups of people can be convinced that they all see the same; this then becomes the "truth" or state of knowledge despite being incorrect. They all accept the story without further questioning or searching for the hidden factor. Help your team members to understand that they need to be critical of published results and conclusions.

The three assignments can help your team members to "feel" the need for critical and independent thinking. It's of great importance to develop and use the "philosophy" in the Doctorate of Philosophy (PhD).[3]

As a next step in the discussion, it's also important to let your new team member see that being independent and critical may actually put their career at risk: if your views are running against the mainstream, it may be hard to get your work published. Therefore, you also need to teach your team members that they should "put themselves in the place" of their most skeptical or threatened peers. Why would these peers be so skeptical? What are their interests? What can you do to avoid having your article or grant application turned down by a skeptical reviewer? Steps to consider include the following:

- ✓ Determine a likely difference between the mainstream and what you and your team members propose.
- ✓ Acknowledge the difference without being dismissive.
- ✓ Suggest that what you're doing is complementary and might lead to a novel strategy; you're standing on the shoulders of giants, including mainstream giants.

---

[3] Bosch 2018.

- ✓ Invite a scientist from the mainstream to give feedback, or invite him or her to participate in your project and compare the old way with the new way.

- ✓ Don't overly use terms such as "groundbreaking," "revolutionizing," and "paradigm switch."

- ✓ Rather you should let the facts speak for themselves: give preliminary or other evidence that supports the merits of your idea so that the idea becomes plausible in the eyes of the reviewers.

- ✓ Some (top) journals don't go by the facts or evidence but just estimate what it will do for their journal impact factor. Submit elsewhere.

- ✓ If you are at the stage of writing a grant application, give a concrete "plan B" for a risky "plan A." List your earlier achievements to prove that you have typically gone beyond the state of the art.

Being honest and accountable

For junior people, it may not be easy to understand what it takes to be an honest and accountable researcher. Even senior researchers lack statistical skills and find it hard to report mistakes they discovered in their earlier work or to openly admit they were wrong when someone else found a mistake. But the reputation of you, your group members, and science at large is at stake, so honesty and accountability are musts.

- **Misconduct.** Universities are strongly regulated around misconduct – e.g., manipulating, falsifying, or fabricating data – which can lead to suspension, dismissal, and/or prosecution. Your articles will be retracted and your reputation severely damaged, and you may be featured on retractionwatch.com or gain unwanted attention from the media.

- **Gray area.** Between conscious misconduct and true scientific integrity lies a large gray zone of unconscious and subconscious misconduct and unprofessional behavior. You may be sensitive to status or financial interests and therefore tempted to somewhat oversell the positive aspects of your work and minimize its limitations. Unfortunately, universities cannot make you disclose status interests, nor financial interests that may be paid to you in the future.

- **Third-party interests.** Researchers often work with third parties under the umbrella of a consortium agreement (funded by a public body or a private-public partnership) or a contract agreement (third party pays for the research). A third party may have strong interests in the outcome of your research, which it needs to support its political proposals or ideas or to support its business: your results should prove to customers that the company's product outperforms that of its competitors. Your relationship with the third party may be discontinued (no more funding) if you report results that conflict with its business interests. Such funding parties often have the right to see your draft articles before they are submitted for publication, and they can pressure you to modify the draft article or not publish it at all.

So you need to discuss with your team the mechanisms that could lead you to the wrong side, and you need to reinforce strict requirements to unambiguously stay on the right side. For example:

✓ Don't let third parties influence your research: not the questions you want to study, nor the experimental or theoretical approach you decide to take, nor the results you deliver.

✓ All additional or ancillary activities you and your team members have on top of your normal tasks as university employees should be disclosed – activities such as being an editor of a scientific journal, member of an advisory board of a political party, owner or shareholder of a spinoff company, or a company consultant.

✓ Add an up-to-date and complete list of activities to personal pages on the university website. Specify the relationship (e.g., consultancy or collaboration) and the terms (e.g., paid or unpaid).

✓ Also report any indirect conflicts of interest, e.g., if you or any of your family members hold shares or have other financial interests in the company you're collaborating with. Any appearance of potential conflict of interest should be avoided by everyone on your team.

✓ Promptly and completely disclose all these (potential) conflicts of interest when you submit an article to a scientific journal. These interests will be evaluated by the editor and reviewers and published to inform readers.

Whenever you use data analysis in your research, make sure that other researchers can fully understand and, if they so wish, reproduce your analysis (see more in Section 3.2). Consider these steps:

✓ Define and openly share your data-acquisition and analysis plan before you start the study, and stick to it. Preregister your project and planned paper with a journal if possible.

✓ Create artificial data sets mimicking your real data (e.g., using data simulation or permutation), and analyze them in exactly the same way as you stated in the preregistration for the initial data. Use the same procedures for data cleaning, for finding patterns in the data, and so on. See what the results from these "artificial" data sets turn out to be, and use them for inference in the real data. Are your findings in the real data perhaps not unique because these or more extreme findings also occur in the artificial data sets? State your uncertainty.

- ✓ Reanalyze your real data by using slightly different approaches for data cleaning, for finding patterns in the data, and so on. See what the results from these alternative analyses turn out to be. Are your initial findings perhaps not so trustworthy because they don't occur with other analyses?

- ✓ Search explicitly in all directions for all possible hidden factors and alternative explanations for any unique findings.

- ✓ Provide all the data (including metadata) and analysis tools (e.g., software), and invite team members (and other peers and perhaps nonacademics) who are not involved in the study to interpret the data (first arrange confidentiality in case of data privacy issues). Challenge them to be your devil's advocate.

- ✓ Software may contain bugs, so proper software testing and versioning or an independent and complete reimplementation of the software is needed too.

- ✓ Clearly describe the limitations of your study.

- ✓ If you detect a mistake in your published work, publish a corrigendum or erratum to the article or retract it as quickly as possible.

- ✓ Check other peoples' articles for corrigenda or errata published by the authors and critical reviews or letters published by their peers: would you still build on the data, methods, and conclusions of the original article?

- ✓ Actively and independently reanalyze other peoples' data if your study is based on their data. Check whether you draw the same conclusions.

A YOUNG PROFESSOR'S ANECDOTE

Obituary for a giant

I am not a regular reader of *Nature* or *Science*, but from time to time I browse through some print copies of these journals. And so, by accident, I ran into a full-page obituary of a person whose name I immediately recognized. Several years ago, as a youngster and not knowing about his status, I had contacted this person and a journal editor because I had developed a divergent view on claims made in two of his papers. I agreed that the results in one study were convincingly replicated in a second, but I disagreed about their medical relevance because *he had mistaken a replicable technical artifact for a replicable and medically relevant finding*. The editor and I got a reply, but to my surprise, it was nothing less than an attempt to create fog and enable the editor to ignore my contribution. My comment was not posted online. Later, when I presented this technical artifact as a side issue in a talk I gave at a major conference, a senior researcher in the audience stood up and stressed that *we'd all overlooked this hidden factor and had fooled ourselves by thinking that replication is the "gold standard" for quality*. I was grateful to him for his public comment, but even today, while reading the obituary, I feel sad about how the divergent views of young researchers may be handled by giants and editors.

TRY THIS!

Use the following assignments to discuss good research practices with your team and to stimulate self-reflection by your team members.

Being independent and critical

- People look at an object and claim that it's a square. Others look at the same object and claim it's a circle. Can both observations be true? Does truth exist?
- Are you more a frontrunner than a follower? Provide facts to support your answer.
- Are there any schools of thought that are in conflict in your field of research? If you belong to one such school, then step into the shoes of someone in the other school. What would it take for them to bury the hatchet?
- If 60 percent of the published literature was nonreplicable in your field, how would you read the next scientific article? Make a list of key points to check.

Being honest and accountable

- Do you know your personality weaknesses? For example, to what extent are you prestige oriented, easily jealous of and influenced by other people's success?
- Consider your recent work. Rate on a scale from 0 to 100 percent: to what extent did you turn patterns post hoc into hypotheses, tweak your data, twist your story, exaggerate your findings, hide deviant details, or downplay uncertainties or risks? Explain how you came to your percentage ratings.
- What concrete actions do you take to protect against "fooling" yourself in your own daily research (Figure 1.5)?

Codes of conduct

- Does your university have a code of conduct, good research practice guide, code of ethics, or scientific integrity documents? Discuss these with your team members.

- Consider all your activities and interests: are there any that could or should be considered as "additional" or "ancillary"? Fill out the following table.

Description of activity or interest	
Possible contribution to the academic and business interests of the university	
Possible conflict with academic and business interests of the university	
Time spent on activity	____ hours/week ____ during/outside work
If paid, then clarify the payment agreement	
Is this activity/interest published on your university webpage?	

Dilemma game for scientific integrity

- Do one of the 75 dilemmas tests (e.g., related to collaborating, publishing, reviewing, career, etc.) at www.eur.nl/english/eur/publications/integrity/scientificintegrity/. Challenge your team members to do the test as well, and use it for discussion.

- Play "The Lab," a game with roles in research on how fraud develops at https://ori.hhs.gov/thelab/. You can do this at retreat, and everyone will sit up and start thinking!

1.5

Advance

What makes you a good team leader?

If you are an early-career scientist, you may have started your first independent project. In a couple of years, you may see yourself running several projects in parallel. For sure you will have made (and will still make) mistakes as a leader. Make time for self-education and self-reflection, and hopefully avoid the following pitfalls. You have recruited a PhD candidate, but your prediction about his or her potential growth proved to be wrong (sadly). You have committed yourself to an externally funded work plan, but the postdoc employed on the project complains about your "just do it" directive style and wants to deviate from the plan. You may have collaborated with a colleague, but negotiations about authorship didn't work out to your satisfaction. You have taken too much work on; it was fun, but alarm bells are hinting that your body and/or mind is out of balance. You believe you deserve early promotion, but your dean sticks to the formal rules to your annoyance. Learn through open-minded reflection on your doing, for example, by taking the following steps[4]:

- Join courses and peer discussion groups on academic leadership.
- Find a senior mentor who can serve as a critical sounding board for you and hold a mirror up.

---

[4] In 1983, David Kolb published an experiential learning cycle: experience, reflect, conceptualize new behavior, and experiment with new behavior; see "Further Reading."

- Ask team members for their feedback on your leadership.
- Admit you have weaknesses – perhaps your leadership style is not particularly effective – and be prepared to change.
- Keep your work-life balance sustainable for the long term.
- Develop an antenna for politics and changes in your organization and society.

People can lead in different ways. Figure 1.7 shows four leadership styles, each with two sublevels:

- **Charismatic style.** You are a leader with an inspirational style (you have an appealing vision, can easily persuade others, and are results oriented) or a coaching style (you listen, appreciate,

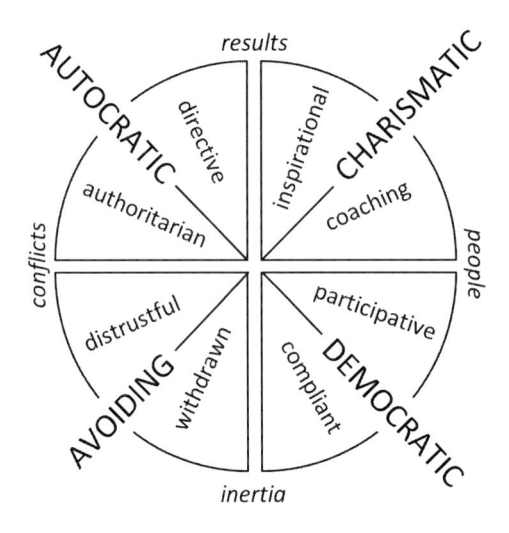

FIGURE 1.7 Leadership styles. There are four main styles, each with two sublevels. When looked at from the outside, do you see an orientation on results ("you hear the workers busy with hammering and sawing"), or an orientation on people ("you hear the team having fun"), or do you notice conflicts between the leader and team ("you hear quarreling and banging doors"), or inertia rather than action ("you hear a desolate silence"). Reflect on your own style!

Source: Modified from Redeker et al. 2014.

and stimulate others, are people-oriented, and look for win-win situations).

- **Democratic style.** You are a leader with a participative style (you include people in processes, accept their propositions, and are people-oriented) or compliant style (you hesitate to give guidance, prefer to stay in the background, and go along with people's interests).

- **Avoiding style.** You are a leader with a withdrawn style (you're absent, don't take up your responsibility, and keep out of conflicts as long as you can) or distrustful style (you don't trust others' motives, think negatively about others, and don't connect with them).

- **Autocratic style.** You are a leader with an authoritarian style (you're harsh on your people, force them to obey, you're not open to criticism, and you go for win-lose situations) or a directive style (you plan, do, evaluate, and act; you go for results; and people have to follow your instructions).

To advance your team and be an effective team leader in academia, you need to be predominantly results and people oriented; i.e., you need to be a charismatic leader and occasionally use a more directive or participative style. The other styles, from compliant to authoritarian, can become counterproductive, if not disastrous, and will not facilitate a group of individuals making an effective research team.

A group is not a team

Two or more people make a research group: Bachelor's and Master's students, PhD candidates, and postdocs, they are in your group and can be busy with their next paper, their thesis, or curriculum vitae (CV). They all run, bike, or skate their own race, focusing on their own goals and successes. But two or more individuals can also work as a team, leading to better papers, better theses, stronger CVs, and better project outcomes, even if they are officially working on different projects. Now they run, bike, and skate wearing the same team outfit and aim jointly to raise everyone on the team to a higher level of success. A sports team includes a physical trainer, sports psychologist, masseur, technician, team captain, and many others; on a research team, you'll have a research assistant, an administrative assistant, and other support staff, with you as the team leader.

Teams typically go through several phases before they perform really well (Figure 1.8).

- **Forming phase.** Everyone is happy with the new job or project; work can start. You as the leader inform, direct, and instruct people about the project aims, tasks, deliverables, and milestones, and you fuel the team spirit and ambition by making the endeavor fun and exciting, something really special (inspirational and/or directive leadership style).

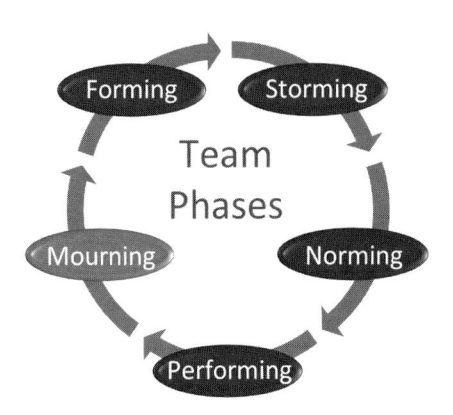

FIGURE 1.8 The five team phases
Source: Tuckman 1965.

- **Storming phase.** There are different opinions or confusion on how to proceed with the details of the work. You coach the team forward (coaching leadership style).

- **Norming phase.** The members negotiate, compromise, convince, or otherwise organize themselves and (re-)organize the work to be done. You enable a constructive controversy and reflection to happen (coaching and/or participative leadership style).

- **Performing phase.** The team and project are alive and kicking. Results exceed initial expectations. You oversee the team and its work (participative leadership style).

- **Mourning phase.** As a project closes, one or more members leave, successes are celebrated, and failures are acknowledged and transformed into lessons learned. You thank everyone for their commitment (participative leadership style).

Then you continue with new projects, each with their own cycle of forming-storming-norming-performing-mourning. See also chapter 3 of the author's book, *Developing a Talent for Science* (Cambridge University Press, 2011). The entire group may go through the stages at the same time, but things may become tricky when different members are actually at different stages because they entered the group at different times. Achieving or maintaining high performance in the midst of many changes of people can be a challenge that constantly demands your attention.

The storming phase may be particularly alarming for new team leaders (and is still challenging for experienced team leaders). Help! What's happening? Suddenly members appear to disagree strongly, show disappointment or anger, and disconnect or revolt. Look at it this way. It's a sign that team members are serious about their work. After all, who would make a buzz about something unimportant, but the storming and norming phases challenge your interpersonal leadership skills even more than the other phases. Some leaders may be tempted to become angry and tell members to obey them (authoritarian style). Others may be tempted to keep out of the conflict as long as they can (avoiding style). Neither style will work well in the long run. Only when you successfully coach the members through the storm will you have established an effective team where the members:

- Have trust and confidence in each other
- Exchange ideas and contributions
- Give and receive constructive feedback
- Go the extra mile for each other and the team
- Express wants and worries openly
- Share fun, enthusiasm, and a high team morale.

The lives of PhD candidates and postdocs can be hectic or problematic for all kinds of professional or personal reasons. They are often in a busy phase of their lives: stormy season. Keep an open ear and eye for their needs and worries, help them to get through difficult times, and arrange a buddy, mentor, health coach, or other internal or external help for them, preferably at an early stage before problems arise or escalate. It's important that group members feel that they can approach you when a problem is arising.

The final phase in a project's life cycle is the mourning phase (also called the "adjourning phase"): the project is coming to a close, the work has been done, and team members will have to leave. In the worst case, former team members have no job to move on to and become unemployed, and you have no funding for the next project and lose your position as team leader. However, the mourning phase can have a much happier end if you take the advancement of your team members seriously from day one – and if you take your own advancement seriously too. Table 1.3 provides an overview of more good practices.

You should also coach team members to work with and support other teams – in the best interest of your team, other teams, and

TABLE 1.3 Leading your team: more good practices in each of the five team phases

Forming
Establish appropriate work conditions for your team, e.g., silent rooms for those who are easily distracted by noise.
Share your vision and work plan for the new project, and invite group members to share their views.
Align objectives of members with the team's objectives.
Foster team interaction by having frequent formal and informal meetings.
Keep your office door open and walk around.
Be a leader who is always prepared to do some of the practical work.
Be curious about what members from other cultures think and do.
Storming
See conflict as a learning opportunity, and handle it with confidence.
Help the team turn mistakes into lessons learned.
Be a critical friend; provide constructive feedback and support when necessary.
Stay committed even if development is slower or more difficult than expected.
Provide training where skills are insufficient.
Norming
What you say and write is what you mean.
Be honest; have one open agenda and no "hidden" agendas.
Treat team members with trust, respect, and pride, even those who prove to have less potential or to fit not so well with the team.

Performing
Share all you know; applaud when team members know more than you.
Reward good team interaction with celebrations (e.g., coffee and cakes), joint papers, thanks, and more.
Make life at work a joyful and exciting experience.
Monitor work-life balance, try to prevent burnout or boredom.
Monitor time, money and quality of work. Act when needed.
Mourning
Help develop team members' future careers beyond your project.
Allow curiosity-driven side activities to become the beginning of a research line of their own.
Help them to leave your team well, even if you will miss them greatly.
Organize a final event for the project team to celebrate personal and team achievements and to close the project in a good way on the personal level.

the organization. Recognize the five phases and be particularly alert to team members speaking negatively about another research team or support department; take action to provide a bridge across teams and jointly enter the performing phase.

Advancing your team members and their careers

Although you may think that running a funded project is all about time, money, and quality of results, the funding agencies also value the *personal and professional development* of PhD candidates and postdoctoral researchers as an important asset of your project. And the funding agencies will evaluate your training success. For example, reviewers for the European Research Council (ERC) are asked:

"To what extent has the principal investigator demonstrated sound leadership in the training and advancement of young scientists?"

This evaluation should be interpreted as your career progression depends on theirs: PhD candidates and postdoctoral researchers are not just a workforce to help you achieve your project's scientific goals. You are their role model and partner on an important journey: their final educational steps to having a fully professional career within or outside academia with a healthy work-life balance. Help them become aware of their goals in their work and life.

Is doing a PhD project a good investment? Will the next postdoctoral period serve their career goals? Or is it time to leave academia, since further training will not add to their chances of achieving these goals? Might it even reduce them?

Outside academia, where the majority of people will eventually find a career, there are many different opportunities: in industry, government, the public sector, or, still close to academia, university administration, scientific publishing, or being a media and public relations officer for a university or research institute. You empower your team members with up-to-date scientific knowledge, with the core skills for doing research, but also with many general skills that are transferable to other settings (see Table 1.6 in TRY THIS! exercise). By contrast, *in academia*, their career journey is toward scientific independence, and you help them develop a research line of their own – not a copy of yours, but something where they may be outperforming you and go on to develop their own opportunities in the academic job market. You empower them with cutting-edge scientific knowledge and the core skills to use it well.

Some examples of how you can help your PhD candidates and postdocs advance include the following:

- **Side projects.** Although they've been recruited for and paid by a specific project, try to set aside some free time for "playing" – just as Google allows its employees to spend a day per week on curiosity-driven activities. It may turn into a new research line for them, and it may also open new or surprising angles for the project on which they are working.

- **Other role models.** Invite guests from within and outside academia to talk to or even work with your team: for example, your alumni (your former PhD graduates and postdocs), your academic and industrial collaborators, an editor of an influential scientific journal, or an officer of a prominent funding agency. Other people can help your team members to sharpen their vision of the future. You can also encourage team members to spend some time outside your group during the project to broaden their experience; for example, they can seek out and apply for travel and internship scholarships and thereby gain valuable experience for future grant applications.

- **Job or grant applications.** Any job application – inside or outside academia – is a quest for money: your team members need to earn a salary in their next job. They may want even more money if they have ideas and plans for the new job that require financial investment by the employer (e.g., for new equipment or support staff). Job competition can be fierce, so teach your people how to make a strong case – a proposal that the other party can't refuse. It makes good sense to teach them or send them to a course on how to write a convincing personal research grant application because this is a skill that is also important for those who want to move outside academia. It shows they can create a vision of where they want to go, outline a concrete work plan and impact plan, calculate the budget, and convince others that such a project is feasible and the investment worthwhile. As the group leader, you will need to help your team members with their job or grant applications. Yes, your investment (time and expertise) will indeed go toward benefiting someone else, but a former group member can extend or strengthen your network by making a new or revitalized

- **Recommendation letters.** Once your team members start applying for jobs, you will need to write letters of recommendation for them (Table 1.4). Pay a lot of attention to these letters; selection panels will read them carefully and consider them seriously. What facts or anecdotes can you share to provide sufficient evidence for your recommendation? You may be asked whether the candidate is among the top 20 percent of his or her peers or to state whether he or she is "very good," "excellent," or "outstanding." Be aware that your personal reputation is also at stake if you are not honest and your statements incomplete. If you don't believe that your team member would be a good candidate for a particular post/job, tell him or her carefully but directly in a positive way and, at the same time, discuss or indicate what kind of job or career track you envision for that person. You don't want to land a colleague elsewhere with a PhD researcher or postdoc who is not suited for the job vacancy. You may decline to write a letter of recommendation, but you should explain your reasons to the candidate, and if you do write one, you should fully commit your time and effort to enable the person to stand out from the crowd.

Your own advancement

Find out what is required for a next step in your career – a tenure-track position or tenure, promotion from assistant to associate professor, or promotion from associate to full professor. Are you expected to take on additional administrative tasks to help your department, faculty, or university prosper? Are you expected to excel not only in research but also in teaching, knowledge transfer, and public engagement, or are you allowed to specialize? What do you actually want? Take time for self-reflection and candor, and perhaps seek help from an experienced mentor. Develop your strategy and action

TABLE 1.4 Letter of recommendation for an academic job (similar for an internship or award nomination, etc.)

Brief introduction
Name of professor or investigator to whom the letter is addressed. Name of applicant and what job they are applying for.
How well and how long have you known the applicant?
A short historical overview, including information on dates and topic of the project the applicant was working on with you.
Science
Mention applicant's main intellectual achievements, publications, presentations, public outreach, other academic activities, recognition, vision, and more (see above).
Teaching
Mention any experience gained in teaching and supervision of students.
Personality
Mention strong skills – expert and general ones. Don't hide any serious concerns you have, but do take into account the candidate's personal privacy (i.e., do not disclose any information you obtained in confidence).
Summary and future
Summarize the information provided in a short but complete evaluation.
You can convert the summary into how the applicant ranks among his or her peers.
Your vision on the match between the applicant and the job vacancy.
Your final recommendation.

Note: Add facts and anecdotes (as proof). Do not show it to the applicant, but submit it directly to the person who requested your opinion.

plan, and discuss your future with your supervisor and administration.

If you want to stay in academia, you may have a large number of working hours ahead of you: 2,000 hours times the number of years until retirement. How can you remain successful and happy for so long? Research work follows a cyclic pattern. You start, run, and complete a project; start again, run again, and complete again, possibly with multiple projects running in parallel. If you successfully revitalize from time to time, you can continue your career in science to the age of 65 or older. Here are some pointers to revitalize yourself:

- Attend conferences outside your field of specialization to see whether ideas that work in another or related field can be introduced in your own field.

- Take a sabbatical of three to six months with another group every six to seven years to get fresh intellectual input or to learn new methods or techniques.

- Negotiate a move from one sector to another within your organization to fuel your research line. Stay there for several years, and then perhaps move on to yet another sector or return to the first institute with your enhanced cross-disciplinary experiences. (This would also challenge institutes to be good places to work in, since a poorly managed institute would not flourish, and people would leave and not return.)

- Write a perspectives paper to outline your vision of where your field should go and get it published. Some hardliners wouldn't consider this as a real publication, but they are wrong, of course.

- Write the textbook you always really wanted to write, introducing students to the results of recent and past research. Usually

- Serve as a mentor to one or more early-career researchers from another institute or faculty.

As a scientist – almost by definition – you are driven by curiosity and maybe at some point you want more, different, or "bigger" adventures in academia. For example, you could want to:

- Combine your current position with a new one elsewhere, e.g., a part-time visiting or honorary professorship at another university.
- Gather many more grants, tens of PhD students, hundreds of articles, many prizes, etc., which could give you an *academic superstar* status. But the mores are changing: the "Matthew effect of cumulative advantage"[5] is considered to disproportionally concentrate resources and reduce the return on funding.
- Take on a limited number of projects only, be there as committed supervisor and collaborator, and let credits for work go to those who really deserve it, which would make you a real *academic leader* and *role model* for others. You go without compromise for high quality instead of large quantity and help change the reward system in science.
- Become a director, assistant dean, dean, vice chancellor, or president of an institute or university or governor of a scientific society or public funding agency or help in the administration of academia. *Beware:* In some quickly developing and highly competitive fields, it may be almost impossible to return to active research.

You may also consider leaving academia – at least for a while or part time – to build a new career elsewhere, for example:

- Combine your current position in academia with a new one outside academia, e.g., a part-time job at a consultancy firm.

[5] Merton 1968.

- Become a full-time researcher or research and development (R&D) manager in industry, start your own business, join a major consultancy firm, or move into politics, perhaps in the department of education.

Leading your team in your organization

It's important to realize that as a team leader you are a middle manager: you are formally responsible for the team under you, but you have other managers above you: a head of department or director, a faculty dean, and the university president. They will monitor and evaluate what you and your team are doing. Let's hope they are content and supportive. But what if they're dissatisfied and impose measures on you and your team? If you agree, you can comply and execute their measures. If you disagree, though, you will enter another storming phase: now it's storming in the hierarchy rather than within the team. It will be particularly tough if the leader above you adopts an authoritarian or distrustful style. As in any other storming event, you can see this as someone who is serious about an issue that is important to them. The negotiation steps you need to undertake include:[6]

1. Don't go into heads-on battle with your adversary. Instead, invest some time in trying to understand what motivates them: pay attention to what they want, need, are concerned about, and their interests. Perhaps they have to cope with political issues or organizational changes you're not yet aware of. Ask, listen, and check whether you've properly understood what they are telling you.

2. Then ask them to listen to your needs and concerns. The other person is above you in the hierarchy, but you don't have to behave as their subordinate. You can remain positive and assertive and, if necessary, indicate that you're also a force to be reckoned with.

[6] Modified from Fisher et al. 1991 and Ury 2007.

3. Refer to a common basis, such as shared principles underlying good research practice. Discuss the results you're supposed to generate, and present the facts to counter any presumptions or misconceptions. Try to work *together* toward a solution that both of you can accept. This might lead to a compromise or a new solution better than either of you could have developed alone.

You and your team will also be working with people who are not in an academic hierarchy but alongside you, for example, support staff such as human resources (HR) and financial affairs. Chapter 2 is about their organizations' rules, processes, and procedures with which you also need to comply. Invest some time in getting to know these colleagues and understanding their work, give them credit when it is due and create a culture of working together. If you take coffee or cake to your meetings and invite them to team parties, it will make it more fun for both parties.

Table 1.5 offers some more tips on how to deal with management and support staff.

TABLE 1.5 Some advice on how to deal with ...

Support staff
Show a real interest in people and their private lives and work.
Place yourself in their shoes; understand their position.
Make them feel part of your team; jointly celebrate success.
Give compliments ("where would we be without you") and credit; help them to understand your position, if necessary.
Management
Be pragmatic with regard to management's rules and procedures.
Be accountable for your actions, including your mistakes.
Learn how to get to a "yes" or to get past a "no."
Stick your neck out when needed, despite possible repercussions for your own position.

AN ALUMNUS'S ANECDOTE

Being nice to people

During my stay at Harvard Medical School, I learned about the American Association for Women in Science (AWIS). A women-only society, this slightly surprised me, but it also made me curious. It was less than a year until my PhD graduation, and the career stories of these women – most of them at a senior stage in their career – could help me make a decision on my next position, whether it should be in science or not.

The obvious step was to sign up for one of their events. But would I dare to talk to these professors? At this point, a lesson from the past came to my mind. My supervisor had once said, "You've all got over 20 years of experience of how to be nice to people, right?," followed by, "Who would like to welcome our world-famous guest speaker and be her host at today's program?" I'd offered to be her host for the day, which had proved to be a remarkable experience, a tipping point in my life.

Thinking about this, I convinced myself to join AWIS and walk up to and talk to professors. Looking back, I met many ambitious professors, as well as other postdocs and PhD candidates from a wide range of nationalities. They all enthusiastically elaborated on their scientific track and shared the lessons they had learned. These AWIS conversations contributed to my orientation process and consequently to my current position in the consulting business. Being nice to people can help lead you to new horizons.

TRY THIS!

The following assignments will help you reflect on your own leadership style, recognize the phases your team is in or has gone through, and allow you to contribute to the training and advancement of current and former team members.

Interpersonal leadership styles

- Could a young researcher be(come) a good team leader? Make your reasons explicit.
- Could someone with no expert knowledge be a good or better team leader? Make your reasons explicit.
- Leaders use different interpersonal leadership styles. Can you give concrete examples from your own experience of each style? How did they feel?
- Which is your preferred style of leadership?
- Ask others what they see as your dominant style, and discuss how it feels for them.
- In what circumstances would you use other styles and why?
- Do you see any reasons to change your leadership style in general or in specific situations?
- A green traffic light means "go ahead," red means "stop," and orange means "be careful." Use green, orange, and red to color each of the eight interpersonal leadership styles in Figure 1.7. Explain your choice of color for each leadership style.

Team phases

- Do you have a group or a team?
- Which phase is your group/team in at the moment?
- How did you react to storming phases in the past?
- How will you react to the next storming phase?
- Have you completed earlier projects in a good manner?

Your current team members

- Do your team members have sufficient time and freedom to develop their own niche in science?

- PhD candidates and postdocs often discount, underestimate, or overlook the value of core transferable skills and could therefore better sell themselves at job interviews than they actually do. Get your team members to list their most important selling points, and see how many of the attitudes and skills listed in Table 1.6 they include. Set up a laboratory meeting to do this or go on an away-day with your team once or twice a year.

- Help your team members prepare for job interviews. Analyze the job advertisement for selection criteria (skills, qualities) and help the team member prepare STARR (situation, tasks, actions, results, and reflect) stories that would inspire and convince the selection committee (see Section 1.3 for more on STARR). For the main criteria, they should develop recent and concrete stories with a positive ending along the lines of the STARR method.

Your former team members

- Where are your former team members now, and how are they doing? Build up a database of them with the following information: name; current and past affiliations and job titles;

TABLE 1.6 Core transferable skills

> Think critically and independently.
> Be honest and accountable.
> Demonstrate strong passion and drive.
> Prioritize, decide, do, evaluate, and persevere.
> Collect, classify, and process relevant information.
> Ask questions for clarification.
> Create ideas to solve problems.
> Analyze qualitative and quantitative data.
> Use current computer software.
> Work on a team.
> Speak and write about a topic to convince peers and the general public.
> Write project proposals, funding applications, and reports.
> Educate and inspire students and others.
> Develop and teach courses.

prestigious grants, awards, etc. for those who stayed in academia; and other relevant parameters for those who moved outside academia. You can add this information to your team website and even to your own CV.

- Are you connected to all of them via LinkedIn or other social media?

- Are there any opportunities to benefit from former group members? Who could you build a new academic collaboration with? Who could suggest candidates for your job openings? Who works at a profit or nonprofit organization that could become your sponsor?

2

"Eighty-five percent of the reasons for failure are deficiencies in the systems and process rather than the employee. The role of management is to change the process rather than badgering individuals to do better."

W. Edwards Deming (1900–1993)
Professor, statistician, management consultant, inventor of the PDCA (plan, do, check, act) cycle

Organization

2.1 Introduction

Many senior scientists look back at the period they spent as a PhD student or postdoc as when they were able to devote almost all of their time to reading new scientific literature, going wild with novel scientific ideas, designing clever approaches to test intriguing hypotheses, developing new theories and collecting strong supporting information, writing and presenting one or more influential papers, going to exciting scientific workshops and meetings, and so on. It was a great and demanding time spent working on important scientific stuff, even though their future career was insecure.

But then your career progresses and your time schedule gradually change. Now it involves writing a grant proposal to raise funding for your first PhD student ("I think I am ready to start my own group") and talking with financial people to budget the personnel costs ("Why does a postdoc cost so much?") to figuring out which other "mystery" experts should be consulted ("Ah, my department head has to approve the application, but only the president of my university can sign legal contracts"). After one or more attempts, hurray, your grant is awarded. But this marks the end of your earlier life as a PhD candidate or postdoc. From now on you are 100 percent responsible for all kinds of matters: recruiting new team members, evaluating and maybe even firing people, keeping track of your income and expenditure, negotiating new contracts, considering activities to generate research impact through outreach projects to the public, or helping a team member in starting a commercial spinoff. Whether you like it or not, you need to learn about the formal management procedures and processes related to human resources (HR), finance, intellectual property (IP), legal contracts, and licenses. The following four sections cover four different administrative or business areas:

- **Human resources.** Hiring a group member, evaluating the performance of the group and its individual members, go/no-go contract decisions, a person ending his or her contract: what are the formal constraints for these and other aspects of HR management?

- **Financial affairs.** What types of costs need to be accounted for? How do you plan the budget before the start of a project and manage expenditures during its duration? Can the project generate any leads for new income streams?

- **Legal affairs.** What types of legal contracts or agreements are you involved in when working at a university and being funded by external parties? What are your rights and obligations? Are your ideas and work legally protected? Who owns what?

- **Patent affairs.** When will companies want to invest in you or your ideas? Or do you want to start your own company? Do you need to patent an application or idea? Can you still publish articles and present your work at conferences? What's the difference in value between a patent and a published article?

2.2 Human Resources

There are many formal processes and procedures for scouting, selecting, employing, training, evaluating, and eventually letting your team members move on. For many of these steps, you can rely on support from the staff in your HR department: HR advisors and policy officers, trainers and counselors, administration services, and more. What can they do for you?

- Discuss recruitment strategy
- Help with position descriptions
- Handle job advertisements and where to place them
- Train selection panels
- Monitor inclusion criteria
- Participate in interviews
- Check qualifications/diplomas
- Prepare job contracts
- Negotiate salaries
- Help with moving in new personnel
- Train you in performance interviews

- Train you in professional and personal skills
- Coach on career choices and job applications
- Advise in cases of conflict
- Implement measures to protect peoples' privacy and data.

But some steps you cannot delegate to others, although support from HR remains crucial. These steps include:

- Choosing between recruiting a PhD candidate and a postdoc
- Specifying the research requirements for a job advertisement
- Selecting the best candidates
- Training them in core scientific skills
- Doing the performance/progress interviews
- Helping your group members with planning their personal and career development
- Deciding about go/no-go points during probation periods.

Several of the human aspects in the preceding lists were addressed in Chapter 1. This section addresses the more formal HR processes so that you can manage your resources properly.

General data protection regulations

As the team leader, you obtain and process other peoples' personal data, e.g., application letters, recommendation letters, employment contracts, appraisal interview reports, and/or information related to illness. Legislation places great responsibility on you and your university in the way that the personal data of students, applicants, and employees are processed and managed, with major penalties for breaches. HR will have designed and implemented measures relating to all its processes – discuss with HR what *you* need to do so that you and your university comply with the rules.

Recruitment policies

You, the university, the funding agency, and the candidates all have strong interests that need to be taken care of professionally. Before you can advertise a vacancy, the legal and financial implications need to be checked and approved (signed off on by those responsible, maybe even by the president of your university). The advertisement and hiring procedure need to meet your university's standards; e.g., is it open, efficient, transparent, supportive, and internationally comparable in its field? Your university or institute may also need to comply with national or internal rules; e.g., if it carries the European Union HR logo of excellence (see Box 2.1 for some of the EU policies that should be adopted in this case). All candidates should be treated equally. This may seem hard or not very practical, but a recruitment decision based on having interviewed one final-stage candidate in person and another by Skype only should be postponed until you have met both in person. There may be additional rules, guidelines, and recommendations to comply with, such as:

- **PhD graduates should leave.** PhD graduates need to broaden their scope, so retaining them as a postdoc in your group may seem good for you but is bad for their career progression. The recommendation may be "no internal recruitment of PhD graduates for postdoc positions." Special circumstances may warrant exceptions.

- **No lifelong postdocs.** Postdocs who have several consecutive fixed-term contracts are likely to end up unemployed with poor career prospects. Maybe there's a guideline that after a couple of years postdocs should either be promoted to more senior positions or leave. Your organization may not allow more than two fixed-term postdoc contracts or set a maximum of four years of postdoc work in total.

- **Internal candidates.** The general rule may be open, transparent, and merit-based recruitment, but in some instances internal candidates may have priority for job vacancies. For example, a vacancy for a research assistant may need to be first offered to employees who are currently, or about to become, unemployed (because their previous project has ended), or this may be used as a mechanism to stimulate

BOX 2.1 *Example of recruitment policy*

Advertisement policies

- Providing clear and transparent information on the whole selection process, including selection criteria and an indicative timetable.
- Posting a clear and concise job advertisement with links to detailed information on, for example, required competencies and duties, working conditions, entitlements, training opportunities, career development, gender equality policies, and so on.
- Ensuring that the required levels of qualifications and competencies are in line with the needs of the position and not set as a barrier to entry (e.g., restrictive and/or unnecessary qualifications).
- Considering the inclusion of explicit, proactive elements for underrepresented groups.
- Keeping the administrative burden for the candidate (proof of qualifications, translations of diplomas, number of copies required, etc.) to a minimum.
- Reviewing, where appropriate, the institutional policy on languages: is knowledge of the national language a requirement or an asset for a particular position?

> **The selection committee should be appropriately diverse**
> - A minimum of three members.
> - Gender balance (e.g., at least a 1:2 gender ratio on the committee).
> - Include external experts from outside the group.

Source: Modified from the Working Group of the Steering Group of Human Resources Management in the European Research Area 2015.

internal talent exchange for a certain category of staff. You may be allowed to recruit externally only if there is no internal applicant or if none of the internal candidates meet the job requirements. This is yet another reason to find out about such requirements because if an employee believes that he or she meets all the criteria, it may lead to an official complaint or even a court case.

- **Recommendation.** A minimum of three letters of recommendation may be required for final candidates. You may be tempted to make your favorite candidate an offer immediately, but there may be hidden factors that you only discover too late (see Table 2.1).

Performance policies

Most universities have annual performance or evaluation interviews conducted by supervisors with their PhD candidate or postdoc and often supported by an HR officer. The meeting has a formal status (i.e., the supervisor may have to grade the performance), and the report of the meeting will be signed by all parties and archived for legal purposes. If each new PhD student or postdoc starts with a clear work plan in the first month, and if you regularly meet and discuss progress with members of your group, there should be no surprises; everyone

85 HUMAN RESOURCES

TABLE 2.1 Example of the mandatory HR processes for recruiting a PhD student or postdoc

	Start project: advertise
1	Obtain formal approval for hiring (signoff by bosses)
2	Write job advertisement (use HR template)
3	Agree on allowance/salary grade (min-max)
4	Develop attraction plan (where/when to advertise, etc.)
5	Develop selection plan/tools (interview, etc.)
6	Arrange selection panel (names, meetings)
7	Advertise
	Start project: select
1	Filter candidates (diploma check, etc.)
2	Short-list candidates (selection panel)
3	Request references/recommendation letters
4	Interview/test short-list candidates (selection panel)
5	Rank short-list candidates (selection panel)
6	Make formal offer to selected candidate
7	Onboard the new PhD or postdoc (all formal aspects)

Note: Color indicates who is primarily responsible: you (black shading), HR officer (gray shading).

knows the project is going well and that the meeting will be a relaxed one. It may focus on:

- **Project strategy.** The progress of the project and its academic output should be evaluated in relation to the original plans. Obviously, at the end of a PhD project, there should be a decent PhD thesis, several of the thesis chapters may already have been submitted for publication, and results will have been presented at workshops and conferences, possibly complemented by some media attention or other forms of dissemination. But since most projects go beyond the original plans due to increased insight during the course of the project, it is essential to refine – or sometimes even redefine – the scientific tasks on a yearly basis or more often. If time permits, some side projects may be initiated (e.g., a follow-up study on some striking or surprising observations).

- **Skills-development strategy.** You can reflect on professional and personal skills: which of the individual's technical and personal skills need extra attention for them to be more successful in the project, and which are important for a successful career after this project. Develop a strategy and concrete implementation plan; take into consideration massive open online courses (MOOCs), traditional classes, coaching, training, mentoring, internships, and similar (see Table 2.2).

PhD candidates are your trainees at the beginning of their contracts and need to grow professionally and personally

TABLE 2.2 Examples of some mandatory HR processes during the running of a project

Run project	
1	New PhD student or postdocs starts work
2	Match newcomer to buddy who helps them feel at home
3	Arrange special working conditions (e.g., disability)
4	Performance evaluations and reports (probation)
5	Performance evaluations and reports (regular)
6	Contract issues (e.g., extension due to parental leave)
7	Training career perspectives

Note: Color indicates who is primarily responsible: you (black shading), HR officer (gray shading), or both of you (white shading).

into independent researchers by the end of the project. It is in the interests of both the candidate and the supervisor that the project succeeds, but it is also in the interests of both parties (and the team) that the contract is discontinued if growth is insufficient for finishing a thesis within the allocated time. Therefore, it may be appropriate to start with an initial contract for, say, 12 months and then decide whether the individual's progress warrants renewing the contract for another two to three years (depending on local employment regulations, of course). Postdoc researchers often get contracts of one to three years at most. These may still include a probation period of, e.g., a maximum of two months (in the Netherlands). By the end of that period, you have to decide about continuing their contract or not – it should have a very high chance of being right for them and for you. Luckily, in almost all cases, this deadline is passed almost unnoticed because the work is proceeding well.

However, supervisors sometimes run into cases where, in retrospect, a no-go decision would perhaps have been wiser and better for both parties. A prolonged stay as a PhD candidate or postdoc may have turned into an unsuccessful struggle that reflects badly on the individual's future career prospects within and outside academia. It is therefore very important to plan and monitor your team members' professional and personal development so that you can base your decisions on mutually agreed-on performance requirements. If well implemented and also well documented, the PhD student or postdoc and you should be able to agree about the situation, and the facts should show that progress and growth were sufficient (go) or not (no-go). You should provide maximum clarity about the requirements for a go, mention in advance if you have any doubts, and warn about a likely no-go, but do allow some time for improvement and offer enough supervision, support, mentoring, and training. Box 2.2 provides an example of clear communication with advance warning, objectives, targets,

BOX 2.2 *Example letter about an upcoming go/no-go decision and what is required for a go*

Project progress, October 1, 2018

Dear [Albert],

You have now been working as a PhD candidate for nine months, while on an initial contract for 12 months. By January I will need to consider your chances of completing a PhD thesis in four years, and the director and I will decide about renewing your contract for a period of up to four years in total.

I had a meeting with you and our HR officer to evaluate your progress after six months, and we outlined an improvement plan for July to September. Recently we met again to evaluate your progress at nine months. My summary of our meeting is given below.

1. On the positive side [refer to action plan]:
   - [write here what went well]
   - [describe points that improved]
2. On the negative/critical side [refer to action plan]:
   - [what failed]
   - [what did not improve]
3. An action plan for the next three months includes
   - [make actions smart, i.e., specific, measurable, acceptable, realistic, time dependent]
   - [meet on weekly basis to allow for lots of support]

We will evaluate your progress again by December 29 at the latest to reach a definitive go/no-go decision about continuing the project. I would stress that, at the moment, it is still possible that this will be a no-go.

Best wishes,

[Ben]

Signature	Signature	Signature (to Show Who Has Read Text)
Ben	Ann	Albert
Supervisor	Director	PhD candidate

and timeline; of course you should first meet face-to-face and afterwards provide (and archive) the written confirmation. Consult your HR advisor for the formal procedures to follow. Most likely you will only be able to legally opt for a no-go if clear agreements have been made and filed at one or more formal meetings and the conditions that would have to be met for a go have been clearly outlined. After a no-go decision, your HR department should help the individual to develop an alternative career path.

The lives of PhD candidates and postdocs can be hectic for many reasons (see Section 1.5). Some may find it very hard or impossible to share their problems and worries with you. As their supervisor, you have a prime responsibility for ensuring that each team member can develop, which is easier if they are happy, but difficult (if not *very* difficult) if they have hidden problems. You need to create a culture in which people can talk to you before it's too late. You need to listen carefully, ask open questions, check whether you have understood the message properly, ask whether there are more issues, be empathic and use your ears, eyes, and heart, and as much as possible try to step into their shoes. Do recognize that there is always the temptation to project your own history and solutions onto the other person. So be aware of your own biases and prejudices, and refrain from making early conclusions, judgments, or condemnation. You can consider asking other people to help: a senior researcher, an HR advisor, a professional psychologist/specialist, or a confidential advisor. Be ready and open to face yourself in the mirror as well. Perhaps you need to learn to show more empathy or to improve your communication or supervisory skills. In some ways, a failed team member reflects on you and your team. Seek a mentor for yourself, and take some courses – your HR advisor can advise you on options. Continue to learn and hopefully you will do better next time.

After a go decision, various new problems may arise that warrant you taking action (e.g., serious issues with behavior, performance, or misconduct). An HR officer can help you to resolve the issues by using informal procedures, which may include internal or external mentoring, coaching, review, advice, and mediation, with the right to challenge each other's views. These are nearly always preferable to invoking formal procedures.

Training and advancement policies

The career prospects of the PhD candidate and postdoc should be discussed. This may be obligatory for PhD candidates in their final two years, but considering future career options at an earlier stage is recommended. You need to help them realize where their talents can be useful in society because that is where their next job could be found. Discuss the opportunities for jobs within and outside academia and strategies to improve their curriculum vitae (CV) over the course of the PhD period. Do develop a plan with concrete action points. For postdocs, discuss strategies of how to fund their own research line and how they can be implemented. Strategic questions to discuss should also include how they can build a network of potential collaborators in their field or across other fields to better position themselves for opportunities. Which smaller or larger national or international grants could they apply for? Which organizations offer the best career prospects?

Give your team members time to explore potential career opportunities, e.g., by participating in a career perspectives exercise (see Table 2.3).

You need to support your group members as they come to the end of their projects, so take the writing of honest recommendation letters very seriously. If there are still loose ends to the project, such as manuscripts that may still need revision, do make agreements about who is going to do what and when. Perhaps you can offer a group member a short-term unpaid contract as a guest researcher (e.g., zero-hours appointment) so that your former group members are not disconnected from everything they may still need (e.g., access to proprietary data and computational infrastructure for revising a manuscript or access to the email addresses they have used for a long time). As team leader, you should ensure that they hand over any data, software, methods, and other valuable items relevant to your group's research and that you (or someone in your group) completely understands what this involves and where to find it (maybe to complete or revise manuscripts). Ideally, the new team member will start the day after the old one finishes or even have some overlap with them. Archive the lab notebooks and any other documents and manuals that contain essential background information for the data, software, and methods. You should also archive all your email correspondence for a couple of years, too.

Keep in contact with team members who leave: you never know whether, when, and how the future may bring you together in mutually beneficial ways. They can become your best ambassadors, collaborators, industrial or philanthropic sponsors, employers of your graduates, or perhaps even your direct boss in the distant future – who knows?

Perhaps you can offer the one leaving a solution to having one or two months' unemployment before their next job starts. If you have some spare money and tasks that need doing, a short-term paid contract may be a good solution. But do check first with HR because there may be legal obstacles (see Table 2.4).

TABLE 2.3 Examples of career perspectives steps for PhD candidates at the University of Groningen

	Prepare for your career as PhD graduate	
Before	PrePhD program to write your own PhD project	
Year 1	Intro to graduate school	
Year 1	Career awareness program	
	Inside academia	**Outside academia**
Year 2	Intro to academic career	Intro to other careers
Year 2	CV and meet scientists	CV and meet alumni
Year 3	Write personal development plan	
Year 3	Academic skills	Other skills
Year 4	Networking and visits	Matchmaking and internships
Year 4	Write grant/job applications	Write job applications
After	Job in academia	Job outside academia

Note: PhD candidates can follow both tracks inside and outside academia to develop their plan A and plan B.

TABLE 2.4 Example of some mandatory HR processes during closing of a project

Close project	
1	Write recommendation letters
2	End of contract issues (unemployment, email access, etc.)
3	Archive relevant material (data, software, etc.)
4	Agree on finishing unfinished business (articles, etc.)
5	Check out
6	Keep in touch

Note: Color indicates who is primarily responsible: you (black) or HR officer (gray).

A YOUNG TEAM LEADER'S ANECDOTE

Paying a high price for a high-impact paper

I obtained a major grant that allowed me to recruit a postdoc for four years. Among the many candidates, a smart young PhD candidate stood out. She was pretty close to finishing her thesis and proactively preparing for her next career step. A phone call with her thesis supervisor, a renowned scientist, confirmed that the PhD work was almost done and that the final version of her thesis would be submitted and approved soon, certainly before she would start working with me. Everything seemed under control, so we signed a postdoc contract. Then we had a wonderful time working together ... ahhh, quite the contrary.

The thesis supervisor wanted her to publish the final part of her thesis in a high-impact journal. And he decided to postpone approval of her thesis until this manuscript was accepted. This required a lot of extra work for her, and she was, by then, my postdoc. This was very stressful for her and very annoying for me and my project. It's practically impossible to have two major jobs to do at the same time. It was almost time for her to look for her next position when she finally graduated with her PhD. We were both unhappy with the lack of success in my project.

I learned my lesson: never again will I hire anyone who hasn't clearly rounded off their previous job obligations.

TRY THIS!

You are part of a large organization, and this assignment challenges you to get to know your organization better in two ways. First, what are the explicit HR rules and processes for you and your team members to adhere to? Second, what are the hidden or implicit HR rules and processes that have shaped your organization and how it behaves now? Check the university internet or intranet for relevant documents, and consult your HR officer to discuss whether you have understood the rules and processes well in all the following assignments.

Find out more about local HR rules and processes for appointing a PhD candidate or postdoc

- How many recommendation letters are required?
- How many staff should be involved in short listing candidates, and how should this be arranged (e.g., independent scoring based on well-defined criteria and electronically stored). Is this followed by a discussion of discrepancies in a panel meeting?
- What is the composition of the interview panel for a PhD or postdoc vacancy? Is one member from a different institute or faculty? Is there at least one male and one female panel member? Are there any other guidelines?
- How long is the probation period for a new position? What steps do you need to take before you can decide on a no-go at the end of a probation period?

Find out more about local HR rules and processes for appointing a research assistant or support staff

- What are the redeployment rules? Can only internal candidates apply, or should they have priority?

Find out more about local HR rules and processes during the life course of a project

- Evaluation interviews: what forms, instructions, and training courses are available?

- Underperforming team member: what documents are there outlining the rights and obligations of the team member and team leader? What steps need to be taken, and who should be involved in the evaluation process?

Find out more about training courses offered by HR

- Leadership-development assessments or courses you can take yourself. Role plays can be very instrumental to spot your areas for improvement, especially if you can see yourself on video.

- Skills and career perspectives courses for your team members.

- Mentoring, coaching, and counseling for you and/or your team members. How can you arrange this and in what instances (e.g., can you arrange a mentor when you become a team leader, and are there counselors to help solve conflicts)?

Find out more about your organization's mission and vision and other HR and HR-related policies that may affect your research

- Discuss general data protection regulations (GDPRs): how should individuals' personal data be managed and processed? What are the penalties for breaches?

- Check for policies on diversity, equality, and inclusion. Check texts, photos, and videos on your team website and your university's website: what can you conclude about the effectiveness of policies on the information they give?

- What measures are in place to foster cross-disciplinary work? What would it take to move your group from one institute/faculty to another (if this would benefit your team's research)?

2.3 Financial Affairs

You, as project leader, write a project plan and specify the personnel, equipment, travel, potential publications or patents, and other items needed to successfully achieve your project's scientific aims and impact. You are held fully responsible for all expenses during the project and especially for not having a negative financial balance at the end of it. You might spend large amounts during your project, at a level much higher than your normal domestic budget, so you may be anxious about being responsible for so much money. The financial officers, project controllers, funding officers, and purchasers at your institute, faculty, or university are the experts to consult for advice and support on direct and indirect costs, the timing of payments, the purchase of goods, liquidity planning, and any other financial matters (see Table 2.5).

Project preparation

Your financial officers will help you calculate the personnel, overhead, and other costs. The project budget shows the total cost and a breakdown into specific items. The financial officers also interpret the funding agency's formal guidelines and can:

- Explain the funding agency's rules on direct and indirect costs and advise you on which costs are eligible to pay from your grant.

TABLE 2.5 Costs that can be directly or indirectly linked to your project

Direct costs	
New personnel	PhD candidate, postdoc
Other personnel	PI (you) and a permanent technician
Travel, hotel, subsistence allowance	In line with usual practices of your university
Goods, services	Lab consumables, supplies, computers
Equipment, small infrastructure, assets	Depreciation, rental, lease, in-kind, paid
Large infrastructure	Operating costs
Dissemination	Editing costs or charges for Open Access publication
Intellectual property protection	Trademarks, registrations, patents
Indirect costs	
Personnel	Replacement for pregnancy leave, extended sick leave
Unemployment	PhD, postdoc still searching for next job; the university may be charged for Social Security allowances
Housing	Office space, organization's desk charges, lab space
Overhead	President and council, administration, PR, teaching services, libraries, etc. "facilities and administration" [F&A]

FINANCIAL AFFAIRS

- Advise you of any costs made before the project starts; such costs will generally not be reimbursed, unless allowed for explicitly in the guidelines.

- Inform you about procurement rules (i.e., when you or your organization can buy goods only via an open-tender procedure so that any external party can make a bid). This often includes the buying of standard materials such as lab consumables or any large-scale supplies you use.

- Help you to obtain quotes from different independent parties for major goods or services you need to buy or lease. You will need to ensure that you get the best value for money. They can help you avoid conflicts of interest and stick to the procurement rules (e.g., not buying goods from a former colleague who is now in business unless you can prove with several quotes that you really got the best value).

- Inform you about local seed money for doing exploratory or pilot studies, making short visits to experts to acquire new knowledge needed in the project, getting strategic and editorial support for writing a professional grant application, etc.

- As a rule of thumb, PhD students are somewhat cheaper to employ than postdocs on an annual basis but not necessarily on an output per euro or dollar basis. And for PhD candidates, you will need to budget for a contract of three or four years. Postdocs are often recruited for one or two years (see Table 2.6).

Universities may receive a substantial financial bonus for training PhD candidates, with a sustainable financial model requiring a university to have 100 PhD graduations per year or more. Indirectly you will benefit, and in some universities your group may even receive a direct share of the PhD bonus (e.g., $5,000 per thesis. Postdocs have more possibilities to find their own funding or a personal grant, so the financial benefit for your group can be large (i.e., on the order of $100,000 or more, unless the postdoc then chooses to move elsewhere with the grant).

TABLE 2.6 Examples of some mandatory financial processes for getting internal approval to start your project

Start granted project	
1	Share formal award letter or signed contract
2	Inform about changes (e.g., delays)
3	Check whether funds are sufficient (e.g., given delays)
4	Cover shortfalls in budget from other resources
5	Open project account
6	Authorize HR to advertise open positions

Note: Color indicates who is primarily responsible: you (black shading) or financial officer (gray shading).

During your project's lifetime

You are responsible for the financial flow during the course of the project. Luckily, the financial officers will do lots of the budgetary work for you. For instance, they may:

- Open an account for managing the finances for your project.
- Instruct you on how to archive all invoices, payroll transactions, or other documents that prove your expenditures.
- Instruct you on how to complete timesheets that prove when and how much time has been spent on the project by you and/or your team members.
- Communicate with the funding agency and book the prefinance, interim, and final payments on your account.
- Pay your project's bills from this account after formal (signed) approval by you and by the project controller who safeguards (stores) the bills and checks their eligibility against the project's guidelines.

- Monitor the expenditure and inform you about timing or if you are spending too little or too much. For example, do not buy a new computer or expensive equipment close to the end of the project because with depreciation over several years, the agency may only partly refund the purchase cost. Often you cannot submit any costs after the project ends; these will not be reimbursed unless explicitly allowed for in the project's contract.

- Prepare the financial paragraphs in interim reports that you (or they) are required to submit to the funding agency.

- Inform you on a regular basis about the financial state of your startup package (if you had one) and of your other projects. They may also inform you about any annual or other financial support given by your institute (e.g., a standard travel budget, bonus for teaching a class, royalties from intellectual property). You will need to have a good overview of your money flow (in and out) and the total balance at all times.

- Charge the legitimate absence of team members to accounts other than the project account, thereby preventing your project from burning up money without making progress, e.g., during parental leave or extended illness or time out to take care of relatives. Unfortunately, some countries still don't have decent maternity, paternity, or care leave (see Table 2.7).

It is likely that you will want or need to make changes to your project plan. Perhaps the field has moved on since you wrote the grant proposal, new technologies have arrived, costs for certain items have changed, or you simply discovered better ways to achieve your overall aim – research projects are to a certain extent unpredictable and therefore funding agencies are generally open to you making changes to the original plan. Thus, they are also open to changes in the budget. Although it is very unlikely that the budget will be increased, you are often

TABLE 2.7 Examples of some mandatory financial processes while running and closing the project

Run project	
1	Archive proofs of all financial transactions
2	Fill out timesheets
3	Report changes (e.g., intention of project extension)
4	Rebook major absence costs (e.g., parental leave)
5	Monitor expenditure against budget and advise/alert
6	Submit financial reports and invoices to funding agency

Close project	
1	Prepare final financial statement
2	Confirm all costs on final financial statement
3	Submit final financial statement to funding agency
4	Resolve residual balance issues (shortage or surplus)
5	Close project once final payments received
6	Assist financial (external) auditors
7	Answer queries from auditors and funding agency

Note: Color indicates who is primarily responsible: you (black shading), financial officer (gray shading), or both of you (white shading).

permitted to move money within or sometimes even between cost categories – but do first ask the funding agency for formal permission.

It goes without saying that you should use your funding wisely and efficiently. The funding agency may be paying hotel and travel expenses for visits to other labs or to conferences, but do not claim anything related to a subsequent holiday. Separate your business expenses from personal ones. Unethical use of any research funding can lead to trouble.

Raising Additional Money

Although the contract with the funding agency has been signed and the project budget has been accepted, there may still be ways to "gain" some extra money:

- If you buy essential equipment that will be used only part time or for less than the depreciation time, then the costs may be only partly covered by the funding agency. But you can offer use of such equipment to other parties for payment.

- Project results can be protected and licensed or sold to other parties, with the permission of the funding agency. This can generate a one-time or continuous cash flow, e.g., from royalties (see Section 2.5 for several examples from the humanities to the exact sciences). Any royalties should, of course, be paid into your research account because they arise from your work (but check your university's rules about sharing revenue). In fact, even your articles and books are protected by copyright and licensed to the publishers, who may pay you royalties for books or chapters you have written (see Section 3.2).

- You can apply to other agencies for small grants to arrange workshops, or you can stimulate your team members to apply for their own travel stipends. This can save some project money that you may be allowed to spend on other activities. You should also put team members forward for prizes, which can include an amount that will cover a conference visit as well as looking good on their CV.

- If your salary is covered in part or completely by the grant, then you are saving your university money if you were already paid by them, say as a tenure-track assistant professor. You should try to negotiate your share of the money saved and keep it in a separate account from the grant so that you can spend it freely, even after the grant period has finished.

- Funding agencies often allow those who are awarded a major personal grant to transfer it to another institution where the conditions are more favorable, in terms of facilities, presence of top peers, or better financial support through a substantial startup budget. Once you have a major grant, other universities may compete for you or try to seduce you by offering you more attractive (financial) conditions than your existing institution. Such offers can be used to bargain with your home institution to gain yourself more financial support. It's a reality check for the ambitious. But your current and potential future employers may see your lack of loyalty as a negative.

Financial barriers

Finances also bring various risks. A grant may cover the direct costs for a PhD candidate but not the overhead costs; the funding agency can simply argue that these overhead costs (e.g., the salaries of administrative staff or the cost of office space) would have existed without your project. But your institute may charge these overheads to you, and consequently your project funding will be depleted more quickly than you imagined. A grant may also cover the direct costs for a PhD candidate for two years only. You may need to apply for another grant to be able to complete your project.

Finances can create barriers between different institutes if they are each financially independent entities, making cross-disciplinary collaboration more difficult. If your government were to pay a bonus per graduated PhD candidate, who would be given the bonus: the university, your faculty, your institute, or perhaps your own research group? Such bonuses can be substantial and should be negotiated at the start of a new PhD project. Cross-disciplinary collaboration may also prove problematic if the funding agency only pays for PhD candidates and postdocs and not permanent staff (after all, they are already paid by the university). If a permanent staff member of one institute spends time (e.g., as a supervisor) on a project that is formally managed by another institute at the same university, the personnel costs may be passed on to that institute. While the project is neutral at the aggregate level of the university (the university pays the supervisors), one institute will now have funding in hand, while the other has a financial deficit. Such short-term win-lose situations may frustrate collaboration between different institutes or faculties and can lead to a lose-lose situation, with fewer grant proposals for cross-disciplinary projects being initiated and the university as a whole obtaining less funding than in a win-win situation. You are leading your team in your organization and must negotiate the best conditions for your team *and* for your university. See the end of Section 1.5 for the negotiation steps.

Financial value

Who are the users and end users of the results arising from your project? Why do they need or want your results? These are essential questions to ask yourself before, during, and after the project. Step into the shoes of the users and end users and try to estimate the financial value of your results for them. Should you still share your results for free, or should you opt for payment? If you do a consultancy project for a commercial company, should only your hours be charged or should you claim a share of the revenues generated from using your input? Perhaps you wrote some software: can you sell it to multiple parties and generate a stream of income so that you can hire someone to maintain the software and hardware, support users and end users, and further develop the software (which can be in the interests of all the purchasers)? Consider the need for further agreements (Section 2.4) and/or protecting commercial interests (Section 2.5). See also the following anecdote.

A POSTDOC'S ANECDOTE

Champagne, but not for me

A startup company is located on the same campus as my office. The company does pretty cool things. One day I met one of the directors, and we talked about a bottleneck in their new technology. I thought I could potentially write some software to solve the issue, and I was happy and proud to receive an offer to work with them – my hours would be paid to the university on the basis of a simple consultancy contract that was quickly signed. And yes, with just 10 hours of work, I solved their problem. My five years of building up specialized knowledge on so-called linear finite mixture models and my experience in implementing these models in software were exactly what the company needed. This was the breakthrough the startup needed to move on to marketing its technology vigorously. The director told me how happy the company was and showed me the champagne bottles the staff had emptied to celebrate this milestone. While being proud of my work, I started to realize how naive my director and I had been: no champagne for us, only a small payment to cover my 10 hours of work, and no share of the huge revenues the company was now generating. On top of this, the company imposed a ban on my publishing the ideas behind the breakthrough. This won't happen to me a second time.

TRY THIS!

You are responsible for your project's budgets and expenditures and don't want to risk problems here. But do you really know which expenses are eligible and which are not according to the rules and regulations of your university and of the specific agency funding your project? Find the correct answers to all the following questions (and consult your financial officer for assistance).

- **Champagne.** You celebrate the project's newest and great publication with one or two bottles of champagne. But who should pay for this: you, the funding agency, or your university? How about the food and drinks served at your Friday afternoon cocktail sessions or at a group retreat?

- **Booking a flight and hotel.** Your team member is going to a conference abroad, which has been specified in the project plan and budget. You can simply book the flight and hotel yourself? Or do you need to follow a formal authorization procedure and/or use a particular travel agency to ensure that the costs are refunded? Can you charge any daily subsistence costs (e.g., meals, taxis, hotel, etc.) or only up to a certain limit? For example, would a first-class train ticket be covered or a business-class flight?

- **Buying a laptop.** You want to buy a new laptop for your team member at a supplier who has a special offer. Can you just go ahead and order it? Find out what the purchasing rules are at your university. And are there procurement rules?

- **Buying expensive equipment.** Your project is moving forward successfully, and after two of its four years' duration, you have the material ready for further analysis, for which you now need equipment that will cost $50,000. Will the funding agency still cover these costs completely? Its depreciation time might be four years, so will you then end up with a financial deficit of $25,000?

- **Printing posters, editing and publishing papers, filing patents.** Posters and open-access papers are standard output from your project, and they can cost you a lot of money to produce. Can you charge them to your project? What about all the costs associated with protecting your intellectual property (IP) by filing for a patent?

- **Overtime.** Your group member works 60 hours a week. Can you charge all his or her hours to the project? What happens if the labor contract between the university and the individual states only 40 hours a week?

- **Work-related hours.** Can you charge the time you spent in preparing the project (before its official start)? Can you charge to the project grant the time spent in preparing meetings, for visiting a general rather than specific scientific conference, for traveling, for work phone calls taken during traveling, and for weekends when certain experiments needed to be monitored or executed?

- **Two or more projects.** If a group member works on two or more projects, do you have to fill out timesheets, and if so, to what level of detail (time units), how frequently should they be checked and signed off – and by whom?

- **Changing internal policies.** What happens if the university changes its internal rates for indirect or direct costs during the course of your project? If this leads to a deficit, who will cover it?

- **Changing currencies.** Your project budget was calculated and approved for one or more years on the basis of estimated costs. But what should you do in the following cases: unfortunately, you now need to buy goods that have become much more expensive due to a less favorable exchange rate. Or what if your grant amount suffers from changes in the exchange rate because your funding agency is based in another country?

2.4

Legal Affairs

If you work with people, you may simply trust their blue eyes or friendly face and work and share your bright ideas and much more with them. And it may all go pretty well. But, just as in a marriage, a signed agreement may be needed when things don't work out well. Because you are probably working with different people and organizations, you will soon have a range of different agreements (Figure 2.1 and Table 2.8).

Preparing for agreements is work for specialist legal affairs officers, who will guide you through the steps from advance notice to the final signing of the agreement (Table 2.9).

A common element in all these agreements is confidentiality and the ownership of existing and new intellectual property (IP; see Box 2.3). What are the implications for you and all the people you work with?

1. **You and the university.** The most important work-related agreement you have signed is the "labor agreement" between you and your employer, the university: your "job contract." It describes the rights and obligations of the employee and employer. It's a right (and a privilege) to enjoy academic freedom to work on your own ideas and benefit from the buildings, laboratories, information technology (IT) infrastructure, peer interaction, support staff, students, management, and the university's branding and reputation. However, you also have an obligation to properly protect the rights associated with your ideas in the interests of your university: you must first keep new ideas and other valuable

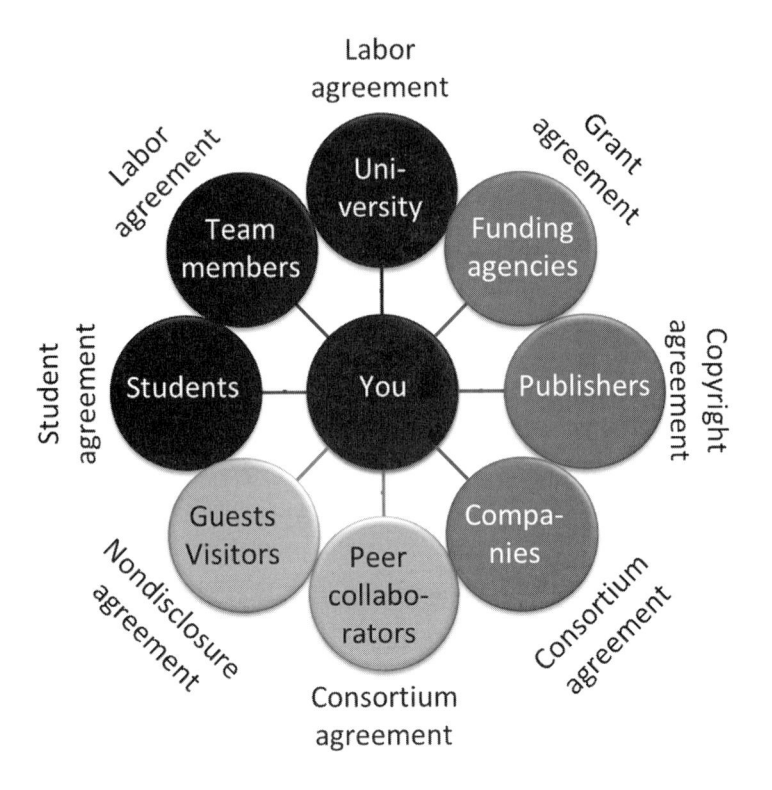

FIGURE 2.1 Examples of the different parties you work with and the types of agreements needed to clarify intentions, expectations, rights and obligations: university-based partners (black), peer-based partners (light gray), and third parties (gray)

information confidential and impress this on your whole team; second, inform and report to the relevant offices; and third, transfer your IP rights to the university. Read your labor agreement carefully to see what else you have signed up for! What if you are also a visiting fellow or honorary professor at another university or if you have two or more part-time jobs at different universities or at a university and a commercial company? Your ideas may result from the synergy between any two environments, which may have set out different obligations, procedures to adhere to, and rules to comply with. Ideally, the two parties should have settled these details in advance by drawing up a supplement to your two labor agreements.

111 LEGAL AFFAIRS

TABLE 2.8 An overview of different types of agreements and the clauses they contain

Generic clauses	
Parties to the agreement	Be crystal clear about who is involved and who is not.
Duration	When the agreement starts and when it ends.
Law	The country whose laws apply to the agreement.
Noncompliance to the agreement	Administrative and financial penalties.
Disbanding agreement	Circumstances under which the agreement can be disbanded.

Confidentiality or nondisclosure agreement (NDA)	
The type of information to be held confidential	This can cover almost anything. Describe the information as completely, accurately, and precisely as possible.
Permissible disclosure of confidential information by receiver	E.g., if the same or similar information becomes lawfully available through other sources or via public channels.
Use of confidential information by the receiver	Be explicit about what the receiver can use and how, when, and for how long.
Rigor of protecting confidentiality	Who needs and can receive the information? Should all parties sign the NDA? How should this information be stored?

Grant agreement (includes all of the above)	
Research plan	Full description of the work, timeline, and anticipated results.
Exploitation plan	Protecting intellectual property (IP); funding agency has a right to object to transfers or licensing to other parties.
Dissemination plan	Are publications and data to be open access? Will there be a time lag in submission of papers so that IP can be protected?
Monitoring progress	Send scientific and financial reports, inform funding body about any circumstances affecting agreement.
Financial compensation	Defines eligible costs, reimbursement rates, payment schedule; rules for purchasing goods, works, or services.
Portability of grant	What happens if you change institutions?
Crediting the funding agency	Precise description of how and when to credit the agency.
Consortium agreement (includes all of the above)	
Division of work	Roles and tasks of parties described as carefully as possible.
Preexisting IP	List of IP that will be made available to the consortium or not.
New (joint) IP	Outline the ownership and access rights. If appropriate, delay publication to protect IP registration.
Governance	Management bodies, their responsibilities and voting rules; how to settle disputes.

Community	How and when to communicate and meet up; what information, forms, and documents need to be submitted and when.
Finances	Division, timing, and distribution of funding.
Letter of intent, memorandum of understanding (MoU)	
Shared interests	All parties have explicitly stated an interest in preparing for or performing a joint activity.
Continuation or disbandment	A statement outlining the circumstances that would lead to either action.

Note: All agreements start with a number of generic clauses. Confidentiality agreements go on to describe how to treat valuable information. Grant agreements describe what the funding agency offers you and expects from you in return. These typically include confidentiality clauses. Consortium agreements describe how different parties are to work together, e.g., if you are working with peers from other universities on a shared goal (often funded by a grant where one party is the coordinator or principal and other parties are the coworkers).

2. **You and the funding agency.** The funding agency is investing in you and your work. It will specify how it wants you to deliver a return on its investment. The agency may want to know your plan for protecting and/or exploiting your IP, so it will want to have a full and clear statement on the IP made available to the project (preexisting IP) versus that generated during the project (new IP). The agency may also want to be co-owner of and/or have a share in potential revenues and/or have the right to object to certain exploitation (e.g., licensing your results exclusively to one party).

We're all used to keeping our research ideas confidential to a certain extent so that we don't compromise our future publications. Sharing a bright idea too early and openly at a conference may allow one of your competitors to legitimately embrace it and perhaps even scoop you by publishing

TABLE 2.9 Examples of mandatory processes for getting a project agreement signed by funding body

Prepare contract	
1	Provide advance notice.
2	Provide contract details (funding body, parties, IP, etc.).
3	Internal bodies approve (finance department, etc.).
4	Assess parties (legal, financial, ethical issues, etc.).
5	Advise on contract requirements.
6	Draft contract or comment on draft contract received.
7	Negotiate and revise contract.
Sign contract	
1	Legal representative approves (e.g., dean, president).
2	Parties sign supplements (e.g., host agreement).
3	University's legal representative signs.
4	Complete and submit signed documents.
5	Funding body signs.

Note: Color indicates who is primarily responsible: you (black shading) or legal officer (gray shading).

BOX 2.3 Intellectual property (IP)

> IP includes know-how, ideas, concepts, inventions, improvements, products, texts, photos, images, videos, music, other artwork, software, databases, access rights, trade or funding secrets, other confidential information, and so on, and it may be protected by copyright, trademark, or patent.

a paper before you can; ideas are free and cannot be protected. But this is a tricky situation because discussing your ideas with colleagues helps you to better work out the details and allows feedback to be given on your ideas. Sharing any details, even a minor one, may render the opportunities for patenting the results originating from your bright idea invalid. It would be too bad to lose publications and patents because of not thinking ahead or preparing (see Section 2.5 for more on patents).

No matter how strongly you believe in or are pressured toward transparency, open research, open data, and open access, issues of confidentiality often remain pivotal and are necessary for publication, as well as legal, commercial, medical, or ethical reasons (such as protecting the privacy of human beings) or to address public concerns (e.g., hazardous biomaterials that can be made into weapons). You should also note that unpublished data, draft project proposals, review reports of papers and proposals, evaluation reports of your group's performance, strategy documents of your group, department, or university, and the names and letters of job applicants should all be considered to be confidential.

The fact that you and your university have signed a labor agreement (a public services contract) and the grant agreement with your funding agency is not the end of the matter. There are more people involved in your research.

3. **You and your team members.** You have recruited group members for one or more projects. If they are employees like you, they will have also signed a job contract. Your team may be running several projects in parallel, and researchers from one project will learn about the progress on other projects during group meetings and when they review each other's draft manuscripts, etc. It may be hard for them to see where and when confidentiality is crucial for their own project, let alone for other projects in which they are not formally involved. Did you share the IP section from the project agreement and

discuss it with the whole team? Did you discuss the intent and implications of the IP section of the job contract? It is up to you, as team leader, to raise their awareness and educate them about such issues, not just once but continually.

4. **You and your students.** They will act like a group member for several weeks or months, perform degree or Bachelor's or Master's thesis projects in your group, and join group meetings where the progress of other people's research projects is discussed. Students have rules to stick to, and they probably know these (it's worth checking), but they may not be aware of the specifics. So be clear about confidentiality and ownership and discuss these issues when they join your group. Help them to prepare their reports and presentations, and instruct them about the issues of confidentiality when they meet external people or even when presenting a poster or speaking at external events. Show them the university's policy documents for students, which they must obey. Some students will stay with you for an internship, e.g., during the summer holidays, while they are officially studying at another university, perhaps with different rules from yours. Get them to sign the appropriate (student) confidentiality or nondisclosure agreements.

5. **You and your guests/visitors.** What should be treated as confidential may be even less clear to colleagues from outside your group who join a group or departmental meeting. They may simply get excited about some of your ideas and use or share them elsewhere, perhaps instantly on Twitter or other social media. Scientists from other universities may also visit your group. They may be young and inexperienced or senior and highly esteemed. They may stay for only a day, a week, or several months. Whatever they learn, they can also take away. Be aware of what you and your group members want to share, especially with peers who have no contract and therefore no obligations, unlike you and your group members who have these defined in job contracts and supplements. If needed, ask guests to sign a confidentiality or nondisclosure agreement when they arrive.

6. **You and your peer/company collaborators.** Your team may collaborate with scientists from several universities and/or companies on projects funded by one or more different agencies. Your ideas meet with other people's ideas, and new ideas pop up from the synergy. Project or consortium agreements spell out who brings in what by way of confidential IP (pre-existing IP) and who owns the new ideas (new IP). Learn to read and negotiate project or consortium agreements when leading or coworking on projects.

7. **You and your publishers.** Scientific journals or publishers will specify the conditions that apply to you as the author of an article and to the people who want to access your paper or book. These conditions define the level of openness of your article, and you should review them carefully before submission of the manuscript. Such conditions can be agreed on as a license between you and the journal/publisher. But do check your freedom of operation: your employer or funding agency may accept certain licenses (e.g., a green model for open access; see Section 3.2) but reject others (articles being made available for a fee). Scientific publishers may claim exclusive rights to the final versions of your articles/books (printed or PDF) or full ownership of them. They will receive revenues from selling these and may offer royalties for book sales. Read more about license models in Section 3.2.

Obligation to report ideas

Any creative work, ranging from an outline of your initial idea to any concrete outcome from that idea, should be properly protected by a copyright, trademark, and/or patent (see Section 2.5). You are obliged to report such work to the relevant university offices as soon as you are "reasonably able to conclude that there is a question of such a creative work."[1] They will then assess the nature of your idea or work and may plan,

---

[1] Collective Labour Agreement Dutch Universities; see "Further Reading."

or ask you to plan, further action. For some creative work, the action required is pretty standard (e.g., specify your name and your affiliation correctly on the article, book, or other publication so that you are credited as the creator/author and thereby also raise your university's standing (which is good for you too). Other types of creative work may require more customized action. Perhaps your software can be distributed under license either freely or for a fee, which could be a source of funding for maintenance, user support, or further development. Brand names, domain names, or logos you create can be registered or given a trademark (but this should comply with your university's policy). Last but not least, knowledge and technology transfer officers scout for patent opportunities (as a welcome sources of extra funding), and you should inform and involve them at an early stage. A patent application must mention your affiliation (patents count in university rankings just as normal publications do). But because there are investments and returns involved, there is more to agree on than simply who owns the IP (see Section 2.5).

Transfer of rights

You may see yourself as an entrepreneur developing your own scientific business and thus deserving all the credit and revenues. In reality, you are more like a franchisee within a host university, and the costs of the holding are huge (way more than your salary and the facilities you use). This warrants a transfer of IP rights in whole or in part from the employee to the employer. You will still be identified as the creator or inventor (this is a moral right), and revenues may be divided between you and the university. Typically, the university will first be compensated for the application/registration, maintenance, and protection of IP (e.g., by a patent), and the remaining revenues may be distributed to the inventors (including students), research groups, faculty, and university (e.g., 25 percent for each). Funding agencies and any third parties involved may claim a share, too.

You can reclaim the transferred IP right if the university does not want to make use of it within a reasonable period of time. In this case, you and/or your research group will make the registration and be compensated first for the protection of your IP, while the remaining revenues may be divided between the inventors, research group, faculty, and university using your university's standard distribution model.

Always be clear about confidentiality

Add a very clear "Confidential" or even "Strictly Confidential" watermark to documents (reports, PowerPoints, emails, etc.), or put this in headers or footers (see the email disclaimer in Section 3.5). You can also say that something should be treated as confidential, but such an oral statement should be followed by a written statement within one or two weeks for your message to be legally valid. Saying something is "Confidential" suggests that the information should, at the very least, not go public, but who may actually see and/or use it (inner circle) and who may not (outer circle) remains unclear. "Strictly Confidential" may suggest that information is meant only for you (i.e., only the sender and recipient are in the inner circle), but without further clarification, the recipients may still make their own interpretation. So be very clear what you mean, and inform others appropriately.

Copyright

Copyright protects your written work but *not* the idea expressed in the text. The same is true for photos, videos, and software. Copyright protects your work in two ways:

- You are the only one allowed to publish, share, or reproduce the work (you have an exclusive right to exploit it).
- Should you license other people (e.g., a scientific publisher) to exploit your work, then they (e.g., the publisher and their

readership) should credit you properly, and they cannot change the work in any way that would damage your reputation (you have moral rights; they have moral obligations).

The copyright claim "made by you" holds only if the work is not trivial and not derived from someone else's work. It should clearly show signs of your intellect, your creativity, or your personal touch. What do you need to do to obtain a copyright in your work? The answer is simple: you only need to indicate your copyright by adding a copyright symbol (©) whenever you think it is appropriate, or "© [your name here]" to mark yourself as the author, or "© [your name and date]" to mark the date you started work on the item. But you do not actually have to add anything because the copyright claim does not depend on the symbol but on you being the "author" of the work.

In the scientific literature, people are allowed to copy (quote or paraphrase) short fragments from your text in their own texts, provided that they credit you properly and reference the original source. Larger fragments, figures, tables, images, photos, or artwork can only be reproduced with your or your publisher's explicit permission, with proper credit given to you and possibly for a fee. Copyright also applies to anything that you put on your website or blog: no one can reuse the content without your permission, but with your consent you, and the source should always be properly acknowledged. In contrast, people are free to add a link to any information on your website because the link is considered an appropriate credit mechanism. You can protest against unsolicited use of your content by accusing people of "plagiarism." This includes content in all the versions of your unpublished articles and project proposals, which fall under the confidentiality of the evaluation procedure; this content is your IP and cannot be shared by anyone (not even by a colleague) with someone else who is outside the confidential procedure of your university and/or at the funding agency.

Trademarks

Brand names, domain names, or logos you create for your products or services can be registered or given a trademark. One can mark almost any name, sound, or appearance under a trademark (annotated by TM for concrete products, etc.) or a servicemark (annotated by SM for services used in the United States and several other countries), but potential confusion with existing trademarks has to be ruled out. Two names may be spelled differently, while the appearance (e.g., font or color) or sound may be considered too similar to an existing item. After official registration, you can use ® as the symbol to indicate your IP. You will need to pay the maintenance fee to the Trademark Office to keep the trademark valid. No one else in your field can capitalize on your success by introducing a similar name in writing, sound, or appearance.

Most names used in science are not registered as trademarks. You simply devise a catchy acronym for your method, tool, or project and put it on your funding proposal or in the title of your article. Unless reviewers complain, the name will then appear in the list of funded projects and published articles. You cannot expect reviewers to do a proper check of registrations or trademarks to see whether the name you chose will cause a problem (e.g., because "ParallelX" appears to be trademarked and you receive a letter from a lawyer with instructions to stop using "Xparallel" as the name for your method because it is too similar). Too bad if you have to change the name after your paper has been published.

In the case of licensing or selling a patented invention to another party, trademark choice and protection are their business. They may want to give your invention a name and maybe even a logo; having a name and logo can help to advertise the product.

A YOUNG TEAM LEADER'S ANECDOTE

The $1 million bonus project

I knew people from the research and development (R&D) department of a leading multinational corporation whom I had met at a select Gordon Research Conference. They liked my work, and we started a small project together, which they later offered to scale up. They suggested a long-term collaboration and would fund a group of four people to work with me on their challenges. We agreed on permission for me to publish, a promise to patent our results, and to protect the company's efforts from competitors, I was not allowed to work for their main competitor. This was an offer I could hardly refuse. But I realized that I was capitalizing on the innovative outcomes of my PhD and postdoc projects. Would this large-scale collaboration really allow me to develop and innovate to a level that I would remain a partner interesting to the company in the future? I shared my concerns with them and suggested I needed an extra $1 million of free money for just this purpose, which would ultimately benefit both parties. They agreed!

TRY THIS!

Discuss with your team

- **Job contracts.** Put the contract on the table, and discuss the sections on rights and obligations relating to IP.

- **Student and visitor agreements.** Check with the legal affairs department whether they offer such agreements, and discuss the sections on rights and obligations with your team.

- **Grant agreements.** Your team is working on one or more funded projects. Discuss *all* the sections of these agreements, one by one, and agree on their implications for your daily work.

- **Existing IP.** Make a list of all your proprietary IP (ideas, knowledge, tools, materials, software, and so on), and estimate its commercial value.

- **Confidential information.** What confidential information about your research would you be willing to present at your next meeting? Check for meeting policies. For example, the Gordon Research Conferences (GRCs) maintain a "no publication policy" (www.grc.org/about/grc-policies-and-legal-disclaimers/) to promote discussions and the free exchange of ideas at the research frontier. Read their clause carefully, and explain your decisions.

2.5

Patent Affairs

The public, through their taxes, pays your salary and funds your project and expects that your ideas and inventions will sooner or later benefit humankind. In return, you should do whatever is reasonable to achieve this sooner rather than later. Some inventions can be applied right away; others aren't ready and need to be further developed by you or by other parties. Such further development and subsequent marketing of the "product" may require huge investments, in which case the filing of a "patent without any prior release of information about your idea" is almost always the way to proceed. Once you can show a proof of concept, patenting opens the way for companies to invest in your product, and this can be the quickest, best, and often only path toward achieving real benefits for humankind.

You can patent technical and process inventions of different types, for example:

- A computer program
- A mathematical algorithm
- A new therapeutic indication or target
- A specific process or route to produce a compound.

Patenting is necessary unless you are absolutely sure that no one can reinvent your innovation (e.g., the recipe for Coca Cola) or because reinventing would be too costly (e.g., completely rewriting extensive software to bypass the copyright on your text).

If granted, a patent will provide you with the necessary protection for your innovation for a maximum of 20 years. This means that in this period you are the only one who can use the invention and that no one else can (commercially) exploit it without your consent. With such protection and your consent, other parties may be willing to work with you and invest in developing your concept into a marketable product. As outlined in the preceding section, you should note that you may be obliged to report your concept to officers in your university as soon as you are "reasonably able to conclude that there is a question of such a creative work." In other words, you may be in trouble if you decide (for whatever reason) that you do not want to consider submitting a patent application, for example, because you think it will prevent you from publishing a scientific article in a high-visibility journal, while the university's patent officer may conclude that you have spoiled the university's opportunity of earning a major amount of money. Equally, the agency funding your project or your partners on collaborative projects might sue you for infringement of the contracts you have signed with them.

What exactly is a patent?

A "patent" is defined as the description of an innovative solution to an existing problem. The patent application describes:

- The problem
- The state of the art in the field
- Your solution to the problem
- Why your solution to the problem is innovative
- Examples of your finding to support commercial applicability
- Exactly what you want to protect, i.e., the "claims" marking out the limits of the invention

- A detailed benchmark to existing products or processes to define the technical or other differences.

Legally, the patent provides the inventor with "exclusive rights" granted by a sovereign state for a "limited period of time" in exchange for detailed public disclosure of an invention. "Exclusive" means that it forbids others to perform commercial activities that fall under the patent. "Rights" means that you can commercialize the invention, e.g., sell products based on the patent or sell licenses for its use. If your patent is built on top of one or more other patents, you will first need to obtain licenses to make use of these inventions if the other patents are held by someone else.

When can you apply for a patent for an innovation?

There are three strict conditions:

1. **It should be new.** The field's state of the art on the "priority date," the day of filing your first patent application for the invention, is what counts. Neither you nor anyone else should ever have shown your invention to someone else or talked about it anywhere (not at a conference or your normal group meeting when external guests were present), nor have you ever published it (not even in a conference abstract, press clipping, blog, or email to an external colleague) before the priority date. If you have to show your idea to people (e.g., to raise funds), then you must make sure that all those present have first signed a confidentiality agreement. New is not the same as innovative.

2. **It should be innovative.** No one with a general knowledge of the field of the invention (i.e., who is supposed to know the state of the art in the field and its literature up to the day of you filing your patent application) should be able to derive your technology or process in a straightforward manner given the field's current state; your invention must not be self-evident by any means. Surprising new combinations can be innovative

and thus eligible for a patent. This aspect is called "novelty." New and innovative are not the same as applicable.

3. **It should be applicable.** Your concept for a product or for a process to produce a product should be concrete enough to be functionally, industrially, and commercially relevant. It may still be in a prototype phase, but a solid proof of concept is needed.

How a patent differs from a normal scientific article

A confidential draft of a future scientific article can form the basis on which a patent attorney can write the patent application (Table 2.10). However, there are some major differences:

- Patents have inventors, not authors, and all inventors must be named on the patent. Anyone else who helped incidentally ("a pair of hands") should not be listed on the patent. Any mistakes in who you name may have major consequences for the validity of the patent.

- You have to define matters exactly in a patent application (see Table 2.11). Your definitions must be precise and complete, whereas you can assume such definitions will be known by the readers of a scientific article.

- You can make claims that you have not yet proven; revisions to the text can be made up to one year after the priority date (see below).

- Claims should be substantial and expected to generate cash within two to three years because a patent application incurs a high fee.

- Every word written in the patent application can have judicial implications. It is therefore crucial to obtain the help of a patent specialist in writing a patent application.

Once your patent application has been filed (see below), you may submit your scientific article to a journal and start talking about your ideas at conferences, for example. Most commercial (project) partners will allow a delay of 6 to 12 months for a patent application to be made.

Filing a patent application

Most universities have a transfer and technology liaison (TTL) office with ties to external patent specialists. They will also help you predict the balance of costs (e.g., $100–$20,000 for the initial stage of application, $10,000 or more for final filing and maintaining, and more than $100,000 for defending the patent) and benefits (e.g., royalties; although net revenues over $100,000 per academic patent are rare, there are incidental

TABLE 2.10 What is written in a patent application?

	Patent	Equivalent in scientific article
1	Inventors	Authors
2	Priority date	Earliest claimable filing date (usually the submission date)
3	Technical field of invention	Field of research
4	Background of invention	Problem stated
5	Prior art	State of the art
6	Summary of invention	Summary of findings/results
7	Description of invention	Detailed description of, e.g., materials and methods
8	Claims	Accurate and detailed description of the scope and breadth of the findings/results/applicability

TABLE 2.11 Patent texts can include many details that may seem irrelevant or trivial to a scientist but necessary to a patent specialist: (a) excerpt from the claims section of a patent on new statistical methods; (b) excerpt from the description of the invention in the same patent

(a) What is claimed	
Claim 3	The method of claim 2, wherein the plant is a species selected from the group consisting of: *Agrostis, Allium, Antirrhinum, Apium, Arabidopsis, Arachis, Asparagus, Atropa, Avena, Bambusa, Brassica, Bromus, Browaalia, Camellia, Cannabis, Capsicum, Cicer, Chenopodium, Chichorium, Citrus, Coffea, Coix, Cucumis, Curcubita, Cynodon, Dactylis, Datura, Daucus, Digitalis, Dioscorea, Elaeis, Eleusine, Festuca, Fragaria, Geranium, Glycine, Helianthus, Heterocallis, Hevea, Hordeum, Hyoscyanus, Ipomoea, Lactuca, Lens, Lilium, Linum, Lolium, Lotus, Lycopersicon, Majorana, Malus, Mangifera, Manihot, Medicago, Nemesia, Nicotiana, Onobrychis, Oryza, Panicum, Pelargonium, Pennisetum, Petunia, Pisum, Phaseolus, Phleum, Poa, Prunus, Ranunculus, Raphanus,* Ribes, *Ricinus, Rubus, Saccharum, Salpiglossis, Secale, Senecio, Setria, Sinapis, Soanum, Sorhum, Stenotaphrum, Theobroma, Trifolium, Trigonella, Triticum, Vicia, Vigna, Vitis,* Zea, the *Olyreae,* and the *Pharoideae.*
(b) Description of the invention	
Soft- and hardware	A set of instructions (embodied in one or more programs) encoding the statistical models of the invention is then executed by the computational device to identify correlations between phenotypic values and haplotypes. Typically, the integrated system also includes **a user input device, such as a keyboard, a mouse, a touchscreen**, or the like, for, e.g., selecting files, retrieving data, etc., and **an output device (e.g., a monitor, a printer, etc.)** for viewing or recovering the product of the statistical analysis.

Source: Beavis WD, Jansen RC. Patents PCT/US2000/034971, WO2001049104A2, WO2001049104A3, priority date 1999/12/30; available at https://patents.google.com/patent/WO2001049104A3/en.

success stories of millions of dollars even in academia). Revenues may be assigned to the funding agency, your university or research group, or even to you personally.

A first step in thinking about a patent application is to check widely and thoroughly: mine the literature and internet, use every search option from PubMed to Google to Google Scholar, and look into patent databases such as Espacenet (see "TRY THIS"). If you find any papers, reports, or patents that describe innovations similar to yours, then the route of patent profiling becomes much less attractive – and you can save the expenditure. From the moment that you decide to involve specialists, it's an expensive business going forward.

The next step is to find and contract a patent specialist, who will do a similar search before even starting to write an application. When the application has been written, the patent specialist will file your application either at your national patent authority such at the Netherlands Patent Office or the Intellectual Property Office (IPO) in the United Kingdom. Your application can also be filed at more "international" authorities such as the European Patent Office (EPO), the US Patent and Trademark Office (USPTO), or the Patent Cooperation Treaty (PCT) for one "worldwide" application that you need to convert later into individual patent applications per country. The day of filing a first patent application on the invention is called the "priority date" – within 12 months subsequent applications can claim the priority date of the first application. The patent application can be filed in any country where you see a market and therefore require a patent, but you need to decide to proceed well before 30 months from the priority date. Your patent application will be reviewed by the national patent authorities (for which an expensive fee is charged). After several rounds of revisions, if you are lucky, your final version will be approved, and the patent will be registered and published (i.e., patent has been granted). Note that a patent may be granted in one country while it is still pending in another.

Also note that anyone can review your patent once it is published and take you to court to dispute the novelty, self-evident nature, or applicability of your concept or product. You will also have to protect your patent, i.e., look for any infringement to your patent and start a court case to call a halt to its use (and you will incur legal fees). Costs for patent maintenance generally increase rapidly after a "low cost" initiation phase of two to three years; if the budget balance is less favorable than anticipated, you can retract the patent by no longer paying the maintenance costs. You should realize that a good termination strategy for your patent is as important as the

TABLE 2.12 Example of what the mandatory patent process may look like at your university

	(a) Patenting process
1	Complete internal invention disclosure form (IDF)
2	Is it patentable: prior art scan?
3	Is it commercially interesting: market analysis?
4	Report to board of university for approval
5	Write patent application
6	File as US patent application (patent pending)
7	Collect more evidence to substantiate the claims
8	Revise patent application (before end of priority year)
9	National patent granted (patent published)
	(b) Develop or terminate the patent
1	File as PCT patent application
2	Lobby to interest commercial parties
3	Negotiate to license or sell
4	Terminate patenting process in case of lack of interest
5	Cash and share royalties and sales in case of success

Note: Color indicates who is primarily responsible: you (black shading), legal officer (gray shading), or external bodies (white shading).

development strategy; otherwise, the costs may rise hugely without you having any prospect of a return on your investment.

Will a patent make your group rich – or poor?

You're a researcher and may have no business or commercial interest. The success or failure of a patent to raise a return on investments nevertheless depends to a large extent on the energy and effort you are willing to put into it. Check some of these requirements:

- You are prepared to visit potential commercial parties, nationally or internationally.
- You can inspire and convince third parties to invest in further (applied) research to develop a prototype into a marketable product.
- You know how to deal with people who are used to being leaders, e.g., your director, dean, president, and investors.
- You are interested in helping to write the patent application and are on top of negotiating licenses or selling the patent.
- You are happy to show up for court cases if the patent is disputed.
- You have a talented, entrepreneurial PhD student or postdoc who is skilled and has the time and ambition to start a business of their own (with you as a great mentor and shareholder).
- You would look forward to designing marketing material, recruiting staff, and hiring external knowledge-transfer specialists (and many more tasks) should you decide to start your own company.

The hard truth is that most patents cost more money than they will ever generate. It is generally assumed that no more than 1:300 or 1:3,000 patents will be profitable in the end. It's all part of the game (while not exactly the same, very few scientific articles are really "successful" either).

Some researchers have adopted strategies and tactics from the business world (e.g., protection based on multiple patents is harder to break than protection based on one patent) and have raised a source of income for themselves, their group, and their university. They can fund their research activities over many years without having to compete for funds from "normal" agencies. Other scientists are named as an inventor on a patent jointly with industry partners, and the nature of the patent is protective; i.e., the claims in the patent protect a broad field in which the company is or might become active (e.g., all crops listed in Table 2.11a). Or the description is so detailed that the use of even obvious tools or methods has been included (e.g., a mouse and keyword to be able to run software, as in Table 2.11b).

A PROFESSOR'S ANECDOTE

Patent and business by accident

I injured my eye in a terrible accident, and it required immediate treatment by a specialist in the ophthalmology department of a university medical center. And so two professors met for the first time, one the patient and the other the clinician. At our next meeting we discussed our research interests, and a surprising interdisciplinary collaboration was born. This is how science can work! We formed a team, the professor of ophthalmology with a postdoc and the professor of biochemistry with a PhD student, and we made great progress, deciding to apply for a patent for our joint innovation and to make a business out of it. The team participated in several venture capital contests. These were very helpful because the postdoc and PhD student learned about the business and economic side of the invention, which is completely different from the scientific side: customers, markets, competition, and how to engineer a basic idea into a real and easily producible product. At each contest we performed better than the last, and finally we won the Dutch Venture Challenge and the Green Talents competition run by the German Federal Ministry of Education and Research (BMBF), and we won an innovation prize from Novartis. The postdoc and my (by then former) PhD candidate started up the business. This is how an accident could be turned into an unexpected patent and new business!

TRY THIS!

You've got an idea

- Check the scientific literature; Google for trend reports, whitepapers, and position papers written for or by ministries of economic affairs. Has anyone anywhere published something that might jeopardize the newness of your invention?

Check for patents in your field of research

- Check patents.google.com to "read the full text of patents from around the world." Insert your keywords.

- Check patent databases, e.g., on worldwide.espacenet.com/advancedSearch, and insert your keywords. Or go to www.epoline.org.

- Esp@cenet contains 60 million patents from many countries, including information about their application status – filed, active, or ended.

- Epoline of the European Union contains all published European and worldwide patent applications, including information on their status.

- Also check www.wipo.int/pct/en/ for all the nations covered by the Patent Cooperation Treaty (PCT).

File your patent in multiple countries

- In which countries do you see a market now or in the future? In the Netherlands, Japan, or China? Motivate your answer.

- Can you file your application in your own language or English in these countries? Or should your application be translated partially or entirely into Dutch, Japanese, or Chinese? Discuss this with your patent officer.

3

"We cannot seek achievement for ourselves and forget about progress and prosperity for our community ... Our ambitions must be broad enough to include the aspirations and needs of others, for their sakes and for our own."

Cesar Chavez (1927–1993)
American labor leader and civil rights activist, major historical icon for the Latino community

Society

3.1

Introduction

Presentations at scientific conferences and publications in journals are traditional ways of sharing your ideas and results with your peers. If you are lucky, many of the relevant people will come to listen to your presentations, meet you at your posters, and read your published papers. They may then start adopting your ideas, using your results, citing your papers, and even inviting you to collaborate in research projects and seeking funding for them. Your personal reputation, or "brand identity," is starting to pay off: your "research business" is growing and may reach higher levels. It's no wonder that publication and presentation lists make up an important part of curriculum vitae (CVs) and play an important role in career progression and promotion.

But there is more to personal branding than using traditional strategies. Communication and collaboration practices are changing rapidly. The internet and social media now allow you to share ideas and results instantly: there's no need to wait for the next conference or for the final paper to be published – and there's an unprecedented audience, not just the limited number of peers attending a conference or reading a certain journal. Anyone with a computer and access to the internet can now read open access material, learn from your ideas, praise them, give critique and contribute further ideas, or add new data that you could not have obtained otherwise. Effectively, you can have access to unsolicited extra brainpower and personnel. Communication can be open (e.g., a blog, open groups on LinkedIn) or closed (e.g., email, closed groups on Facebook). Anything that is open can be counted (e.g., number of downloads, tweets, or likes) and can affect your scientific reputation (both positively and negatively).

These new media can boost the outreach of your research business to unprecedented levels. However, this also applies to your peers and certainly to science in general. Fast knowledge dissemination accelerates validation of recent results and knowledge development, leads to earlier adoption by practitioners and society, and alerts funding agencies and aspiring students and employees to what you are working on. What else could you and the scientific community wish for?

The next four sections cover various ways of reaching out to people in society, from peers to the general public.

- **Open science.** Is there only one dominant flavor of openness, or are there more to consider? Can the "ownership" or "authorship" of your data, texts, figures, and more be protected well when you engage in open access? Should you formally license your work to others, and what do green, gold, and diamond licensing models mean?

- **Citizen science.** Who wants or needs to learn about the results of your research? Could their interest be so strong that they may even want to contribute to your research in some way? Will taxpayers want a say about your research?

- **Media.** What's the key message you want to share? Who is in the audience? Why do they want or need to hear your story? What would be the best medium to use to reach out to them? Can a newspaper or radio journalist become an ambassador for your message? Should you use the power of social media?

- **Web profile.** Everyone uses Google, so what will they find if they look for you or for your topic? Is your personal website popping up first, and does it offer inspiring and complete information for people from peers to the general public? Do you come across well as a team leader? What happens with your personal website and email account if it is maintained by your employer and you move from one employer to the next?

3.2

Open Science

As a scientist, you are generating a gradually growing stream of output, such as:

- Scientific and popular articles
- Monographs and textbooks
- Material for courses
- Instruction and lecture videos
- Raw and processed scientific data
- Software for data analysis and visualization
- Materials, tools, and other concrete products
- New methods, processes, protocols, etc.

Any such outreach that is clearly labeled as yours can boost your research and career in many ways. It is likely to bring you:

- Invitations to speak at conferences
- New, useful collaborators
- People soliciting for jobs at your lab
- Revenue and new funding to continue your research
- Increased chances for promotion to full professor
- And so on ...

Results should spread like oil, quickly and widely, to as many of the potential beneficiaries as possible. This also serves your PhD candidates and postdocs who need items on their CVs to help them obtain a next job. See the anecdote at the end of this section for an example of combining fast blogging with the normally slow article submission.

Copyright licenses

Copyright protects your texts, photos, videos, etc. but not the ideas expressed therein (see Section 2.4). In a copyright license, you specify what other people can do with the resource (Figure 3.1 shows a license logo). Table 3.1 shows five different ways of licensing your work to others. Apart from the zero option, though, they all require that you are recognized and properly acknowledged as the author.

TABLE 3.1 Creative Commons (CC) license options

3.1a The main Creative Commons license options

Zero		
BY	"Made by me"	
SA	Share alike	
NC	Noncommercial	
ND	No derivatives	

Note: CC is a nonprofit organization that has released widely adopted copyright licenses known as "Creative Commons licenses" (www.creativecommons.org).

3.1b Creative Commons license options: what the user is free to do

Zero	To copy, modify, distribute, and perform the work, even for commercial purposes, all without asking permission.
BY or SA	Share – copy and redistribute the material in any medium or format.
	Adapt – remix, transform, and build on the material for any purpose, even commercially.
NC	Share – copy and redistribute the material in any medium or format, but not commercially.
	Adapt – remix, transform, and build on the material, but not commercially.
ND	Share – copy and redistribute the material in any medium or format for any purpose, even commercially.

3.1c Creative Commons license options: the conditions that apply

Zero	When using or citing the work, user should not imply endorsement by the author or the affirmer.
BY	User must give appropriate credit, provide a link to the license, and indicate whether changes were made. User may do so in any reasonable manner, but not in any way that suggests the licensor endorses the user or users' use.
SA	If the user remixes, transforms, or builds on the material, he or she must distribute his or

	her contributions under the same license as the original.
NC	User may not use the material for commercial purposes.
ND	If user remixes, transforms, or builds on the material, user may not distribute the modified material.

FIGURE 3.1 Example of a Creative Commons license logo

You can, and should, check publishers' guidelines – before submitting your manuscript – and analyze their licensing models in detail. But there is also a classification system to help you: the green, gold, and diamond models (although there are more colors in use). See Table 3.2.

Academic articles

We typically submit our papers to conferences and/or scientific or academic journals, and after one or more rounds of peer review, we hope that the paper gets published in a conference proceedings or in a decent scientific journal. It is time for a celebratory coffee and cake once it has been accepted! However, we may be more concerned about the high impact or other status of the conference or journal than about the licensing model used by that same organization.

TABLE 3.2 Access model of journals coupled with their minimum licensing model

Open access model	Review process	License model	Fees to be paid
Green	Maybe	NO license or "BY license" without time delay.	Neither author nor user pays.
Gold	Yes	BY license for final version of the work only, sometimes with serious time delays.	Author or institution pays, but no user fee.
Diamond	Yes	BY license for final version of work only, without time delay.	Neither author nor user pays.

- So ... is the paper now freely available to anyone interested or only to a restricted set of (fee-paying) people?
- Are you, as an author or coauthor, allowed to post your text version (or even the journal's final, copyedited and typeset version) on your own or your university's website?
- Is access open without delay or only after, say, an embargo of a year or more?
- What do you have to pay to gain open access? (There are different levels of this.)

- Is there open access without review, or before, during, or only after (peer) review?

- Are all versions of the manuscript open and/or are the reviews open?

What would be a suitable license model for your scientific papers? Probably you agree it is a license of the type CC-BY (see Table 3.1): peers are free to share and build on your publication, but only if they properly cite your work *and*, if they use CC-BY themselves, should they opt to distribute their derived work. BY requires proper reference of your paper, i.e., names of authors, article title, journal title, year of publication, and page numbers, with possibly a URL or DOI. You may want to add SA, i.e., choose the license option CC-BY-SA, which makes commercial distribution of a derived work harder because buyers of that derived work are free to build on it too (or to copy and redistribute it).

An example of BY in the green model is *arXiv.org*. This is a repository with 1 million prints and preprints in the exact sciences (from physics to quantitative biology to finance). Researchers can post and instantly share their manuscripts at no cost and read and cite manuscripts posted by their peers. *ArXiv* holds the nonexclusive, irrevocable right to distribute all these manuscripts. The green model allows you to post your print or preprint immediately on your own website or in your university's repository (or in commercial repositories such as ResearchGate and AcademiaNet; see Section 3.4) or to submit your preprint to a scientific journal. Some manuscripts shared via *arXiv* have never been published elsewhere and yet become very influential in their field. More recently, other repositories have been launched, e.g., for the biosciences domain (bioRxiv.org) and psychology domain (psyArXiv.com). Some scientific journals even upload all submitted manuscripts automatically to the corresponding Arxiv.org (e.g.,

PLOS submissions appear as preprints on bioRxiv, and comments posted to bioRxiv can be used in the *PLOS* review process). Reviews and manuscript revisions can be made available with open access as well. With version control, the manuscript can become a "living document" that may be improved over time.

Academic books

Books can also be published with open access. Electronic versions can take the form of a single, downloadable PDF file of the entire book, a collection of PDF files (one for each chapter), or an eBook version that can be uploaded to an eReader. Costs for editing, design, marketing, printing, and distributing books can still be significant and might be covered by traditional sales of paperback or hardback copies and sales of eBooks to individuals and university or public libraries. Open access can effectively help market your book: people will share, tweet, or blog that there is a free book out. Even free open access can stimulate sales because some readers and probably more libraries will still want eBooks in their systems or physical copies on their shelves. Whether open access sales would be better than traditional sales channels is a function of the number of people who get to know about your book and the proportion of them who want to buy it. The time and cost to market a book may be considerable, too. Revenues may come in only after quite a delay. Several institutional or national initiatives allow authors and publishers to achieve upfront funding (e.g., oapen.org library). You can consult your librarian for advice. The agency funding your project may also cover the costs needed to produce your book – therefore, you should budget the open access costs in your grant application.

Printed papers and books are static, whereas digital publications are easy to correct, extend, and revise, and it is easy to use social media or email to alert potentially interested people. Digital work can even become a community effort – Wikipedia is a typical example. So these may be additional arguments for you to consider open access publication.

Further considerations are the total publication costs and how these could be reduced by open access; the quality of review processes conducted by open access journals – and the price of that quality; and the risk that a publisher goes bankrupt, in which case access to existing papers (including yours) may cease (websites become unavailable and who knows who owns the copyrights of your papers now?).

Data and metadata

Funding agencies have invested large amounts of money in your project, and although you have probably done great work (published in the relevant journals, obtained patents), their investments are at risk if crucial original and processed data and metainformation on that data are lost when the PhD candidate or postdoc leaves your group for a new job elsewhere. Think of information on the design of experiments, questionnaires, notes in lab notebooks, preprocessing and processing methods, and software in all its versions, for example. Any result from your project that can be digitalized can be stored and shared. Other researchers can redo analyses, but more important, they can combine your information with other data and solve bigger puzzles with the help of big data infrastructure and experts.[1] The agency's investment in your project can have a more lasting and greater effect. With a CC-BY license for access to your information, this may also increase your bibliography or other metrics as a positive side effect. Many researchers believe that open access sharing of data and metadata is an important condition

[1] Trelles et al. 2011; Prins et al. 2015.

for transparency, reproducibility, and integrity in research. Not all data can be shared: a "privacy impact assessment" (PIA) may be needed to ensure compliance with general data-protection regulations (e.g., by pseudonymization). The library and/or information technology (IT) experts from your data management center can help you develop a professional information management plan (see Table 3.3) and implement it.

- **Dataverse.** The Dataverse project (powered by Harvard University) is one example of an open source web application to share, preserve, cite, explore, and analyze research data and metadata (www.dataverse.org). You can install it locally and display it on your own website. But your IT office may arrange for you to have a Dataverse account as part of a larger Dataverse repository. Users of your data can, and should, cite it using the digital object identifier (DOI) for your data.

- **GitHub.** The GitHub project is an example of an open source web application to collaboratively develop research software, maintain its full history (versions), and share it with the outside world. DOIs to your software (versions) can be obtained via www.zenodo.org (powered by CERN Data Centre), which is integrated with GitHub (e.g., http://zenodo.org/record/13200). You can also create a personal website to your repositories (https:/github/yourname).

- **Notebooks.** You can describe all metadata in the old-fashioned way – writing in a bound paper notebook or using a modern electronic notebook or lab journal. Keep track of how your idea and its implementation develop on a daily basis. Notebooks also serve as the archived memory of all relevant research details that may otherwise be easily lost (e.g., due to staff leaving); a skilled reader should be able to fully understand and replicate what has been done. In a paper notebook, you can paste in relevant graphics, photos, and images and add a summary description plus ID numbers and a secure

TABLE 3.3 Elements of a plan for preserving research information digitally for future use by others (first column) and an example showing a more detailed, concrete plan (second column)

General	Example from plant biology
Owner	PI Dr. X; PhD candidate Y
Describe research	DOI for the publication
Study design	
Types of data	Genotypes, phenotypes, software
Experimental design	Randomized block design
Data production methods	Genotyping arrays, carbohydrate profiling
Computational analysis pipeline	Preprocessing, analysis, post-processing methods/software
Data archiving	
Data standard	MIAME
Version	DOI Dataverse, DOI GitHub
Size	100 GB
Metadata	DOI notebook
Costs	Covered by university for first 10 years
Data access	
License model	CC-BY after end of project
Privacy impact assessment (PIA)	No opt out for open access (no data privacy issues)

Note: These elements comply with FAIR (findable, accessible, interoperable, reusable) data principles (Wilkinson et al. 2016).

storage location for any external information that is relevant but cannot be included in a paper notebook (e.g., digital data on a data storage server or biological samples in a population bank freezer). In an e-notebook, you should include accurate links to everything. Other researchers can then replicate your work and combine your information with their own and others for meta-analyses. Notebooks therefore may be crucial for "open research" – to reach further and achieve a meaningful return on (public) investments in past projects.

For free or not for free?

Professional users may benefit from paying for your resources. For example, if you sell them software or access to software, you can potentially hire a software engineer, maintain a professional website and hardware for running the software, offer support and online information on FAQs, develop further releases with new features, and so on. You may want to grant nonacademic users a nonexclusive and nontransferable CC-BY-ND license: they can use the software, make profits, but cannot transfer or sell the software or any derivative. You may grant them the right to use the software "just like a book"; it may be freely moved from one computer to another, but only one person can use it at a time. You may offer more expensive licenses for multiple users. At the same time, you may grant your peers a CC-BY-SA license; they don't pay you and can share derivatives if they use the same licensing conditions as you. Or you can grant them a nonexclusive and nontransferable paid license CC-BY-NC-ND and make your endeavor into a small spinoff business (but first check your job contract and/or university rules about where any proceeds should go).

Patenting may also be an option. You license the patent to one or more other parties for royalties or simply sell it to an industrial or commercial company for a good price. Or you can start your own spinoff company. Unfortunately, the patent application and/or restrictive licensing may (dramatically) slow down the wider uptake of your invention and reduce the return to taxpayers or benefit to humankind. This would be a real pity if your invention could actually be made quickly applicable by other researchers or be useful to industry or other parties without further investments.

With "serving humankind" as a guiding academic principle, you can chose for "open science" as the way to proceed. For example, promptly and openly share your new vaccine against a disease; then it can be manufactured quickly at the lowest cost.

A STATISTICIAN'S ANECDOTE

Tweet speed

"Funding agency discriminates against female researchers." This message was published in a prestigious scientific journal. The media jumped on it with bold headlines, politics, and hot debate. I looked at the paper the day it was published and found it wasn't discrimination, but bad statistics. A letter to the editor would take at least two months to get published – if ever. I decided to write a blog immediately and to tweet to some influential people who would retweet the blog. My blog is public and quickly attracted thousands of readers. But that's not where it ended – media attention in national newspapers, highlights in *Nature* and *Science*, a live interview on BBC radio, and finally, a peer-reviewed letter published in the same prestigious journal 10 weeks after the original article. To date, the story is still running: journalists now call me to comment on the statistics in all kinds of other news. Imagine what my CV section for "public engagement" looks like! All thanks to my writing one blog.

TRY THIS!

- The Creative Commons (CC) website offers a wizard for choosing a license and creating html code to add to your document and/or website: http://creativecommons.org/choose/.

- Use Google Images and filter for those you can use without a fee but with proper credits.

- Check the publisher's copyright and self-archiving policies: www.sherpa.ac.uk/romeo/. Fill in the journal title or publisher's name, and check whether the publisher publishes green and gold journals and books.

- Find your data management office, and ask how it supports data and metadata archiving and sharing.

- Check for data standards and software in your field. What do you need to do to comply with them?

- A figure can tell more than a thousand words. If you have made such a figure, make it citable with a DOI (see Section 3.5), and include the figure in a paper with the DOI in the reference list.

3.3

Citizen Science

Engage citizens

There are seven billion people on Earth, so there must be people who can help you with your research in one way or another – be it in cash, in kind, or otherwise (Figure 3.2). Engaging just a tiny fraction of this huge "crowd" could make a big difference to you and your research. The terms "crowdfunding" and "crowdsourcing" refer to this enormous new potential arising from engaging this extra people resource.

Citizens may contribute

You can beg for a financial donation, and more money will certainly help you run your research or business faster and perhaps better. You can also consider going beyond (passive) donors and donations. Allow people to participate and contribute in other ways: use other people's talents, brains, energy, equipment, personal networks, organizing capacities, ideas, knowledge, solutions, and so on, and surprise yourself and others with results that could not have been achieved otherwise, or at least not at the same pace. Take Wikipedia – we all know this and other examples – where projects were run on a more open basis with remarkable results. Will this be on the horizon for your research, too? Unexpected opportunities may

155 CITIZEN SCIENCE

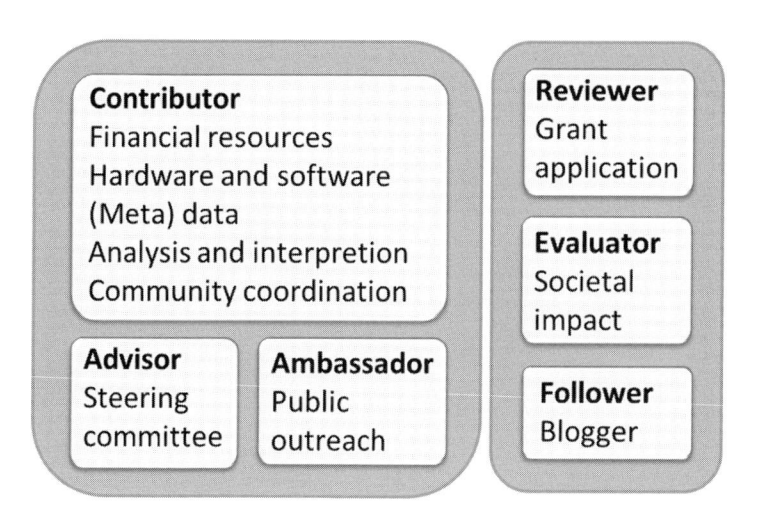

Contributor	Reviewer
Financial resources	Grant application
Hardware and software	
(Meta) data	Evaluator
Analysis and interpretion	Societal impact
Community coordination	

Advisor	Ambassador	Follower
Steering committee	Public outreach	Blogger

FIGURE 3.2 Citizens can play various roles in your projects: some participate in the project (left), while others judge the merits of your project without participating (right).

come your way. If only you knew where to find these potential contributors and how to contact them.

The internet, powerful search engines, and social networking sites have revolutionized access to information and people. Anyone can now initiate a community with similar interests or can connect to a community initiated by someone else. Worldwide, people with access to the internet can read about the community and decide to join in. You can find them too. Search for the key players in that community, and get in touch with them.

People from communities are much more likely to react and contribute than random people from a crowd. For example, if you are researching the long-distance migration of a bird, say the red knot or Artic tern, then local communities of bird watchers or wildlife protection activists may be very interested. Some people live in Finland where the birds breed, others live in China where the birds stop to feed during their migration, and yet others live in New Zealand where the birds retreat during the northern winter (Figure 3.3). These people

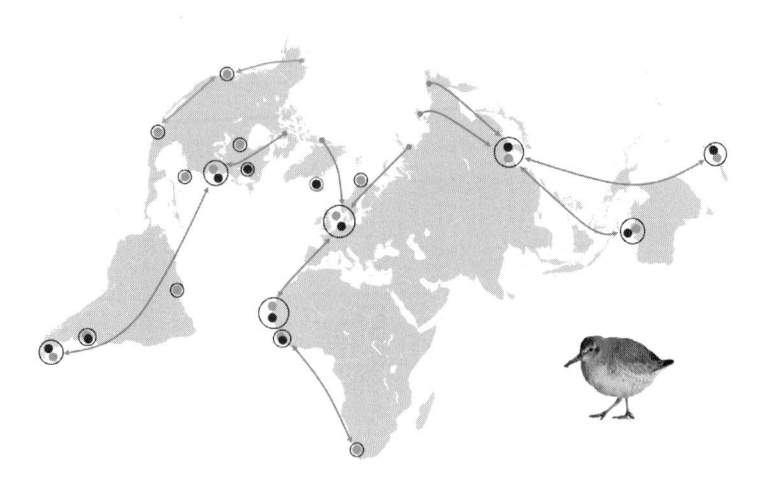

FIGURE 3.3 Citizens help scientists monitor red knots along their global flight paths ("flyways"). Red knots breed in the northern hemisphere and then use different flyways to reach the southern hemisphere. Circles indicate stopover sites for feeding, and black and gray circles indicate locations of two global communities of bird watchers.

may volunteer to record the colored rings on the birds' legs and thus provide important data for you at hardly any cost. They are emotionally attached to these long-distance migrating birds and believe that the birds deserve better protection in a changing world. Being part of your story may have a real value for them (Table 3.4).

Crowdsourcing and crowdfunding initiatives may also be published on dedicated web platforms. For example, some universities already have a website to post and manage crowdfunding initiatives set up by their own researchers (see, e.g., www.rugsteunt.nl for the Ubbo Emmius Fund, University of Groningen, The Netherlands). There are also commercial platforms, and publishing your project on a dedicated platform may help you reach out to large, enthusiastic crowds. For example, Harvard University demonstrated that a computational biology problem could be solved much better, cheaper, and faster via a dedicated platform with 450,000 IT specialists than would have been possible via the traditional way of

TABLE 3.4 Some statistics for the bird watcher example

Citizen group makeup
>7,000 citizens.
3,200 belong to one of 20 organized communities.
3,800 are "loners," individuals not associated to any organization.
>51 countries, covering all continents.

How scientists meet citizens
People walk up to scientists when they're doing field work.
People find websites developed by enthusiastic volunteers who collect and offer data for research (e.g., www.cr-birding.org).
People are members of associated communities (e.g., www.waderstudygroup.org).

How citizens' contributions are appraised
Always reply to all incoming emails: 5,000 per year.
Inform people by email about the life stories of "their" birds.
Send regular newsletters to inform all involved about progress.
Accept invitations to speak at meetings of the communities involved.
Invite top contributors to join you for research expeditions.
Acknowledge their efforts in 200 research and 50 outreach articles.

writing a grant and then recruiting one postdoc: in only two weeks, 122 different individuals submitted viable solutions, 16 of which outperformed existing software; the top five solutions were made available under an open source license afterwards.[2]

---

[2] Cameron 2013.

There are some other platforms that can be used to get something done for a fixed price. The first person to finish the task will win the proclaimed award; all the other people working on the problem are too late, won't make any money, but hopefully enjoyed working on the assignment, learning about the winning solution, and being part of a competition.

Citizens may advise, review, or evaluate

Citizens may donate to patient or special-interest societies and other communities – your research may be funded from these public or semipublic resources. You can invite volunteers from public or semipublic bodies to play a role as an advisor/ambassador for your project. Funding agencies may ask you to include several citizen stakeholders – users and end users of your project outcomes – as members of an official steering committee. The funding agency may also recruit citizen stakeholders as volunteers to provide a nonexpert review (Table 3.5). Decisions on funding will be based on normal peer reviews and these nonexpert reviews. The same citizens may also be asked to evaluate the progress, impact, and outreach of funded projects. Do try to value the experience and expertise of citizens, and certainly don't underestimate them: they can be extremely knowledgeable. Some are professors or retired professionals who may have a specific interest in your project (e.g., because they or their relatives suffer from the disease you're investigating).

Other people who don't participate in or contribute to your project may still follow you and give unsolicited opinions to influence the public debate and political decisions, for example. They may do this by writing blogs or comments on social media, which can be positive but also may be highly critical (see also Section 3.4).

TABLE 3.5 Example criteria for review from a patient perspective. Similar types of criteria can be formulated for other categories of user/end user

Relevance for patients
Does it improve patient health?
Does it account for diversity between patients?
Do patients want or need this?
Participation
Are patients' experiences used in addition to scientific expertise?
Is a patient representative of a community or an official member of the project committee?
Do patients help in developing the research question, plan, etc.?
Information and communication
Will patients and their organizations be fully informed about the results?
Are the communications understandable for patients?
Ethics and safety
Are values and norms of patients respected?
Are all professional codes of conduct, guidelines, and policies satisfied?

Engaged alumni

Your university alumni are a special category of citizens: they already have a bond with your university, their alma mater, and they may feel indebted for the training and advancement that helped them develop their own careers. To show their gratitude, they may be willing to make a donation more readily than nonalumni. For example, a researcher who contributed to an archeological exhibition was allowed to advertise at the exhibition for crowdfunding for her next pilot study. The exhibition had two opening days especially for alumni, during which the crowdfunding target of $15,000 was easily reached. The key to this success was the lectures given by the researcher and the pride of the alumni in being part of this special exhibition and a new research study. In her next grant application, the researcher mentioned the crowdfunding and results of the pilot project – which definitely helped her to obtain this new major grant!

Some alumni may be very happy to share their current expertise and experience for free or to introduce you to their professional network and help you get new commercial collaborators. If they are wealthy, they might offer your university philanthropic funding for specific researchers or for research projects, for which you and/or your team members may be eligible. Do get to know your university's alumni officers, and see what they can do for you.

Develop your strategy

What you need is a marketing strategy and plan – as concrete as possible – outlining how you could get citizens involved. This includes thinking about these four "what" and four "how" questions (4W4H):

- What are the results of your project?
- Who wants or needs these results?
- Why do they want or need these results?
- Where can you find these people?
- How will you present the project and results to them?
- How can they contribute to your project?
- How will you reward them for their contribution?
- How can all of this be organized?

Crowdsourcing and crowdfunding may arise naturally and for a relatively low organizational cost. Nevertheless, you need to deal with the social contract between you and your crowd properly.

- People contribute to a person: you!
- Manage their expectations well.
- Don't promise more than you can really make happen.
- Be open and transparent.
- Involve them in formulating the research questions.
- Share the work plan, timetable, and budget.
- Regularly update them about progress.
- Welcome their perspectives and perceptions.
- Show them that you're listening.
- Be responsible and trustworthy – from the start to the end.
- Reward them by expressing your gratitude – and do more where possible.
- Reward them by participation whenever possible.
- Make the rewards proportional to the contributions.
- Tell them that small contributions are very valuable too.
- Work closely with an influential community member whose advice and reputation will help you maximize your outreach into the community.

Citizens can actively contribute to your science, although there are some issues that you should consider and need to control. For example, some volunteers may deliver low-quality data or fabricate data to get a reward from you. Also, you should clarify any financial issues (will you reimburse specific costs?) and legal issues (confidentiality, copyright, intellectual property, data sharing, privacy protection).

Input of citizens in open science

Open access to the *output* of science projects is necessary for science to be open, but it is not sufficient. There also needs to be an opportunity for *input* from the general public and societal groups. If we haven't listened to them right from the conception of our research agendas to understand what really ticks or bothers them, and if we haven't consulted them during the course of the projects to hear their perspectives, any lack of their applause at the end should not surprise us (see next section for worse reactions). There is even a threat of coming across as arrogant, patronizing, or focusing on personal interests (the academic in his or her ivory tower). Indeed, because of a skewed incentive and award system in academia, questions that are important to the general public may have been ignored over those that yielded higher "internal" academic esteem.[3] Scientists need to earn the esteem and trust of the general public (or research budgets may be cut), show great respect for them, involve them not only at the end but also before the start of new research. This includes engaging with the public, listening to them, framing their issues both rationally and emotionally, finding out what they are really after and what eventual solutions might look like, and getting a two-sided and continuous commitment for our projects.[4]

A PROFESSOR'S ANECDOTE

The moderating mother

I am not on Facebook, and I don't want to be. Let me put it more strongly: I shouldn't be, at least not for my research. Still, my research benefits a lot from Facebook. How come?

As a medical researcher, I see patients. And once in a while a patient with a really rare disease walks into my office. It's unlikely I will ever see a second patient with this disease.

---

[3] Miedema 2018. [4] Maister et al. 2000.

Nevertheless, I want to do all I can to help this individual. For one person with a rare disease, with probably only a few hundred affected (young) people worldwide, Facebook changed everything. It connected my own patient with similar patient and families from all over the world: Brazil, Saudi Arabia, the United States, and further afield. These people had set up a Facebook community themselves. The mother of one of the patients went on to play an extremely important role in setting up my research project: I wrote a short piece about what I could do for such patients, and she advised me on sensitive issues because she knew much better than I did how people would react. She posted the piece about my plan on Facebook, as well as a flyer about my group, my research work, and an email address at which to contact me. What would happen? Would people read it and respond?

The results were astonishing. In no time at all, parents of patients started to email me, sent signed forms of consent and filled-out questionnaires, and delivered other high-quality clinically relevant information. My study went from one patient to 77 in no time, and I had more detailed and useful information than I would typically have obtained directly from their doctors. Suddenly I had a solid basis for doing research on this rare disease. But with so much data, I needed a person to analyze it properly. This is costly, and I wrote a proposal for a grant (which was awarded!). But guess what? Parents of patients started to organize fund-raising marathons and other activities for my research. Wow! So Facebook made my life as a researcher easy?

Well, no. "Easy" is not the right word. This is a serious business for these families. My team and I generate hope in the hearts of these families, and we feel highly responsible for managing their expectations well, for updating them with regular newsletters about progress, for keeping the contact alive for as long as needed by demonstrating that

> we deserve their trust in as many ways as possible. Working with a parent as a moderator between us and the Facebook group was (and is) probably the key to our success.

TRY THIS!

Analysis

- Define types of communities that could show an interest in your research.
- Check websites, Facebook groups, and so on for existing communities.
- Search for peers who may already work with these communities (you don't want to compete unnecessarily).
- Identify key players in each community.
- Define the unique opportunities you can offer each community.
- Define rewards for donations in cash or in kind for each community.
- Make a community-specific research plan including a budget and timeline to share with them.
- Contact the key players and ask them for their opinion and feedback.

Marketing

- Think up a unique selling name for the project, and use it consistently throughout your marketing.
- Set up a project website with an email address, with their names reflecting the name of the project.

- Set up social media accounts: URLs for Facebook, Twitter, YouTube (see Section 3.4 on media).

- Write a short, intriguing, and compelling story about you and your project.

- Make a video (see Section 3.4).

- Write a (regular) newsletter (see Section 3.4).

- Always ask the key people in the communities for feedback before you publish the newsletter. Or let them publish it!

- Consider contacting local newspapers and radio and television stations first, and scale up later to national newspapers or magazines.

- Add URLs to all your email communications, personal website, blogs, your contributions to other people's blogs, and so on.

- Stimulate people to spread your information.

Wrapping up

- Do you now have answers to all the 4W4H questions?

> **What** are the results of your project?
> **Who** wants these results?
> **Why** do they want these results?
> **Where** can you find these people?
> **How** can they contribute to your project?
> **How** will you reward them for their contributions?
> **How** can all this be organized?

Responsible university and team

- Science has been dominated by *white men from Europe and North America*. For example, scientists have traditionally studied health far more in men than in women. They have studied questions that are simply irrelevant to people who take a different view of nature and the environment (such as Native Americans, Aborigines, and the Maori). What is your university doing to counteract such a bias, and what does it expect you to do (e.g., courses for students and staff on inclusiveness)? What can you do to raise your team members' awareness?

- Climate crises, banking crises, biodiversity crises – does your university see it as its obligation to take on the most urgent global challenges? The United Nations adopted goals to end poverty, protect the planet, and ensure prosperity for all: "for the goals to be reached, everyone needs to do their part." Is your university taking responsibility for bringing about change, and what does it expect you and your team to do? In what ways can you educate your team members to be responsible citizens and scientists? See www.un.org/sustainabledevelopment/sustainable-development-goals/.

3.4

Media

The media landscape is changing rapidly. *Traditional media are slow and serve a one-to-many communication channel.* For example, a journalist interviews you and prepares the content for a printed newspaper or radio or TV program. *In contrast, modern social media offer rapid one-to-many and many-to-many communication.* Using web technologies, anyone can become a reporter and prepare, share, add to, revise, or sometimes delete content, and anyone else can join in and express appreciation or dislike or contribute more detail. Of course, mainstream journalism has gone online too and mixes traditional with modern media.

Media connect people. If you don't know the other people well, your message may be ill-phrased or go over their heads. If you don't know yourself well, your message may easily harm your own and your university's reputations. Therefore:

1. **Get to know your target group.** Of course you can't shake hands with all your potential readers and listeners and ask about their knowledge and perception of science in general and of your topic specifically. But it is essential that you know what you are targeting: which national newspaper, popular science magazine, radio program, TV talk show, alumni magazine, secondary school newsletter, or social media network are you dealing with? Take your time, do your homework, and study a couple of issues or broadcasts for format, focus, and style. Go for it if

you think you have a strong story to offer in the language of the target audience – photos and videos included. Identify a number of people from the target audience, and test your message out on them in a pilot study to smooth out any problems.

But there are more people involved than just the target audience, particularly in traditional media, the journalist who will write about you and may interview you is an important person.

2. **Get to know the journalist(s).** In most cases, the journalists will be on your side. They need a good story as much – if not more so – than you do. Consider dropping a journalist a short email or try to call him or her if you have a story to pitch, especially if you have read a nice news item in the science pages of a national newspaper written by this journalist. It may pay off immediately or a bit later. Find the journalist on LinkedIn, search the internet to learn more, find an icebreaker topic, and use it to connect as human beings by asking about the journalist's education, interests, preferences, and contributions to the news media. Once you start to work with a journalist, discuss what he or she is aiming for and assess who else he or she could interview. Ask whether you can check the final report for inaccuracies in content and in your quotes, but be aware that their timelines can be very tight. Be cautious if there is a preference for hype or scandal – don't say anything you don't want to be quoted. Once you have built up a good relationship with a journalist, you can drop him or her an email once in a while if you have an exciting new story. The journalist may also start calling you for your commentary on breaking news in your area of expertise.

You may be a first-timer in "media land," and there is a risk of learning the hard way the "it didn't work out well" lessons. So there is yet another important person who should be involved in your communication: the university's communication and media officer, who is the expert on press communications and/

or has access to a team of in-house or external experts to support you.

3. **Get to know the university communication and media officers.** These officers have built up crucial experience with researchers and journalists, so they can assist you and a journalist to find common ground. Above all, their role is to establish and protect the university's public reputation in the news media. They have a large network of journalists to work with, and some will be favorite journalists whom they know personally. They will also have their favorite news channels and can open doors for you. They will want to write the press releases for you because they know better than you how such pieces should be written. They can pitch you for radio or TV programs that might be looking for scientists with media appeal. Your press officers will offer you tips and tricks and media training for conducting radio and TV interviews. They will use social media to increase attention and know which platforms are most influential. They are extremely important if things "turn rough" or (better) can teach you how to avoid such risks. The press officers are also responsible for building the university's strong branding: the public needs to see where their tax money goes, what sort of research is being carried out, what the societal implications are, and so on. You are just a tiny part of a larger branding effort. Make sure that they have your cell phone number, and be available after a press release or for a media performance.

Dazzled as you may be with adrenaline and the excitement of getting media exposure for your work, you may easily forget that there are yet more people in the communication chain: those who should be credited, have interests to be protected, and are entitled to limit or object to your media performance.

4. **Get to know who else has an interest.** This group definitely includes the publisher of your upcoming article. The publisher may inform the press, and you can prepare for media exposure under embargo until the journal's press release is out. If the

media attention is related to a running or finished project, then you have to check the contract between you and your funding agency: it may specify obligations concerning dissemination. Most likely you will have to acknowledge the agency properly in all your dissemination activities, but maybe you also have to inform them beforehand so that they can also prepare a press release as part of their own branding. If the media attention relates to outreach for a collaborative project, you should also check the consortium agreement for obligations: inform collaborators beforehand, properly credit the consortium and any individuals who contributed. Of course, you should also seriously consider whether you are actually the best person to represent the consortium in the media: don't be selfish, but hand over to the consortium leader or another collaborator if he or she is in a better position to deal with the media. What holds for collaborators also holds for coauthors if the media attention relates to a multiauthor publication. If you are spotted by journalists for developing new intellectual property (IP), you will need to be extra careful not to infringe on ongoing procedures to protect that IP. So ask your legal affairs and/or knowledge transfer officers, check agreements with (industrial/commercial) partners, and together set the boundaries between what can and cannot be disclosed at this point. Also check whether there are any IP issues with photos and/or videos that will be used by the media.

The most important person in communication is the most knowledgeable (in terms of the science), but he or she can be the weakest link in the chain if unprepared or inexperienced in media performances:

5. **Get to know yourself**. The very first question to consider in any request for a media performance is "am I really the right person for this?." Be prepared to say no and kindly decline the invitation if the topic is outside your field of expertise. Never let yourself be enticed into making statements that you cannot back up with scientific literature. You may be tempted to express your personal opinion, but you simply cannot do this

when representing the university! A second question to consider is: "what do I need to do or learn to prepare for this media performance (see Table 3.6)?." Ask for and accept support with writing a "teaser" or a press release. Discuss tips for a one-on-one interview by a journalist, and check what other people's experiences are with the journalist if his or her name is known beforehand. Prepare your key messages, and practice hard for any radio or TV performance. If you are called for a live radio interview, check whether there is time to prepare yourself, and ask them to call back later – or say no. Otherwise, your audience may remember you well – but for something other than the message you wanted to communicate. Finally, check your internet presence. Is it up-to-date? Your tweets, LinkedIn profile, and personal or university webpages may all pop up, especially if the journalist is seeking ways to include stories relating to the human side of your work.

TABLE 3.6 Tips for your media performance

Seven factors for SUCCESS in the media[5]
Story, Unexpected, Credible, Concrete, Enthusiastic, Short, Simple
General tips for dealing with the media
Seek the media only if you have really hot news to share.
Share who did what, when, where, and why (W5).
Avoid jargon, abbreviations, and acronyms.
Use simple analogies or metaphors.
Don't oversell, exaggerate, or hype.
Prepare one key message that everyone can remember.
Prepare 5 to 10 catchy soundbites and rehearse them.

---

[5] Heath and Heath 2007.

Accept that the journalist will filter and rephrase your input heavily.

Ask for questions beforehand so that you can prepare your answers.

Prepare for 10 to 15 questions you don't want to be asked.

Take a media training course.

Tips for a press release

Make a catchy title.

Put the most important story element first.

Use 300 to 500 words max.

You may include quotes from you and photos/artwork.

You should include your full contact information.

You should be available by phone after the release.

Tips for a radio interview

Ask how long it will be; be prepared.

Speak in a lively manner but slowly, and articulate well.

Ask to redo a Q&A if it wasn't strong the first time.

Don't feel pressured by silences; take your time.

If it concerns a debate, avoid getting emotional.

Play the game on your side; stay with your key messages.

Find a quiet place for doing a phone interview.

Extra tips for a TV appearance

Look at the interviewer, not at the camera or ceiling.

Use face and hand expressions (e.g., smile).

Show up well dressed, not over- or underdressed.

Get some coaching beforehand; rehearse with a camera, and check how things come over.

Engaging with journalists

You talk – via the journalist – to your audience (Figure 3.4). Use the journalist's questions as entry points into your story, as starting points to share your key message(s). Play with the questions; take the lead. Your audience should see you as competent, reliable, and credible, and therefore you can share your knowledge, but more important, you can explain its implications and its added or emotional value. Someone listening is looking for "what's in it for me?" or "how is this going to affect me?" Why do they want – or need – to know your story?

A journalist's role is to report on novel developments but also to check whether public money has been well spent. Journalists do their homework and try to uncover any hidden issues. If you need to control damage, for whatever reason, make your story short and dull (repeat the key message + too dull to make it into the wider media).

Stop your reply to a question promptly; avoid droning on. The most important punctuation mark in interviews is therefore the full stop ("period"). Newspaper journalists in

You talk via Journalist to your Audience

FIGURE 3.4 The who, what, and how of communication. Who are they, the people in your audience? Do you know their knowledge level, as well as their interests, opinions, worries, and emotions? Why do they want or need to know or learn from you? What do you want to share with your audience? Do you really have hot news to share? Is there one key message they should remember above all? Don't try to be complete on all the details. How will you inform and engage them? Sharing knowledge and facts will make you look competent, reliable, and credible; showing vision, passion, and empathy and sharing human interest anecdotes about successes and failures and metaphors, pictures, or videos will engage them. How will you play the game with the journalist and make him or her your enthusiastic ambassador?

particular take plenty of time for an interview, and you may become talkative and fail to recognize their dashed questions (Figure 3.5). Examples of dashed questions are the "what if ..." that challenges you to speculate and the "if ... then ..." that is both directive and suggestive. Refuse to answer these questions, but you could offer the journalist a perspective (e.g., the name of a person who could perhaps answer the question), or you can promise to investigate the issue and indicate that the journalist can call you later. If you feel that your answers are going in the wrong direction, simply say, "I'm sorry, that was incorrect, I meant ..." and start again with a different answer.

Anything you say before or after an interview could be used by the journalist; some will continue to audio record or keep a video running after the official interview has finished. There is no "off-the-record" conversation with a journalist; the interview is only over when you or the journalist has left. You can also audio or video record the interview yourself, which can help stop journalists from asking dashed or black (misleading) questions (see Figure 3.5).

Be totally clear to the journalist that you won't accept a press release or news article that is not truthful, complete, and clear. Agree that you can read the article before it is published, and correct phrases that are wrong. Be particularly careful with headlines because they may not be conceived by the journalist but by a different person. You don't want your story to be misread, hyped, or dramatized as media channels are prone to do.[6]

Which social media to use – if any?

Are social media worth the extra time investment when you're already so busy running your research project?

---

[6] Shmerling 2016.

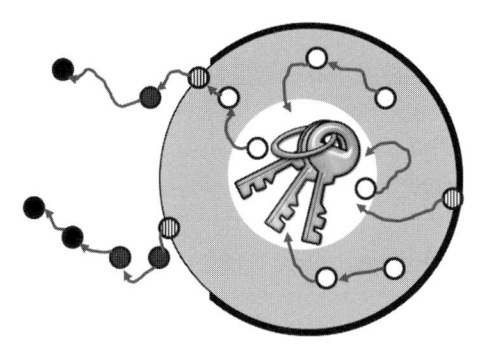

FIGURE 3.5 The interview pitfall. You prepared key messages (inner circle) and know the boundaries of your expertise (outer circle). A journalist asks questions (dots), and you reply (story lines). Be particularly cautious with questions at the boundary (dashed dots): the journalist may challenge you to answer them and subsequently fire off further compromising questions (black dots); you will later seriously regret your answers. Therefore, build a wall so strong that a journalist can't break through it (thick black circle, not partly but completely surrounding the outer circle). Then the interview will be your opportunity to convey your key messages to your target group.

- **You're a nonuser.** You're already overloaded with work. If there was anything exciting to report about your project, the university communication officer would probably prepare and post a message somewhere. You prefer to ignore the social media (hype) and use your precious time wisely. Well, read on; perhaps you will become convinced that it's worth becoming socially engaged.

- **You're a consumer.** You know that your peers, the traditional scientific publishers, and the funding agencies have adopted social media. Automatic alerts help you keep an eye on their messages. You don't post any messages yourself, but you use tools and mine sources that can help you. For example, you're alerted when new papers come out or when a new call for grant proposals opens, and your political antennas are alert for changing governmental or societal trends in science and concerning scientists. Perhaps, from time to time, you could start contributing yourself; see Tables 3.7, 3.8, and 3.9 for suggestions.

- **You're an active contributor.** You're excited by the opportunity to share project progress and receive feedback from anyone who is serious about your topic during all the stages of your project. You use social media to discuss hot topics and recently published articles. You enjoy it and are prepared to invest the extra time to educate the general public. See Table 3.10 for some etiquette rules.

No matter how much you like or dislike the role of social media in academia, it's very likely to play a growing role. The number of times your article has popped up in social media will be counted and converted to a score (e.g., Altmetric) and serve in evaluations of your impact and performance, in job and grant applications, and for promotions, awards, and prestigious memberships (although media attention is no guarantee of quality or impact). It's certainly no longer the traditional "publish or perish" or "get cited or perish" world, based only on scientific articles that are cited in other scientific articles.

Social media can be classified into two main categories: those supporting traditional one-to-many communication but now with you in the role of the reporter and those supporting many-to-many communication in social networks. Some are specialized in only one type of digital information (text, photo, audio, or video), whereas others allow for uploading multiple types. Joining networks of people may require registration, creating a personal profile/account, and acceptance by a moderator, after which you can connect with the group and may receive automatic, profile-based suggestions for people with whom to connect. Social media tools can also be effective in scouting for candidates for your research vacancies (see Section 1.2 and its anecdote). Sadly, social media tools can turn out to be disruptive when sensitive personal data are collected and abused, such as in the recent case of Facebook and the company Cambridge Analytica. General data protection regulations (GDPRs) need to be improved and enforced more than ever before.

TABLE 3.7 Examples of how social media can serve your research business

Microblog (e.g., Twitter)

Stay informed on upcoming calls and other news by following a funding agency's microblog.

Stay informed on opinions of important researchers by following them.

Ask a question or make a comment during a scientific conference or even during a presentation.

Share a newly published article with peers.

Blog (e.g., your Wordpress blog, PubPeer, Wikipedia)

Read postpublication reviews of scientific articles, e.g., to see whether people have reported flaws.

Add your knowledge to a free encyclopedia, and provide links to papers, including your own.

Write and publish your review of a science article or book.

Social network (e.g., Whatsapp, Facebook, Skype)

Tune in with your group members who are dispersed over different hotel/conference rooms or at home/office (e.g., Whatsapp group).

Connect patients to your project, join and share information with their community, and ask for feedback.

Announce on a student network site that you have great Master's thesis projects, and call for candidates.

Discuss project tasks and progress with dispersed collaborators through video conferencing.

Professional networks (e.g., LinkedIn, Google Scholar, ResearchGate)

Before you meet a potential collaborator, see whether they are linked to someone else from your institute.

Follow the careers of your former students and staff.

Help academics to find you and your papers easily.

TABLE 3.8 Tips for active microblogging (e.g., Twitter with 140 characters)

Focus on your niche.

Engage followers by linking to extra content.

Use a URL shortener (i.e., bit.ly/ow.ly) for links.

Put the link halfway through the tweet.

Label relevant words with a #hashtag.

Monitor tweets carrying your #keywords.

Include a person (nonprivate) by adding @username to a tweet.

No more than one to three tweets per day.

Be generous and retweet items of interest.

Check other options, e.g., photos, videos, private (group) tweets.

With a video, you can go way beyond text and images, which opens up many exciting opportunities for dynamically engaging with your target audience (see Table 3.9).

- **Smartphone.** Today's smartphones can generate good-quality videos. Just record and check whether sound (background noise), light (contrast), and stability (wobbling) are okay, and if not, remake the video. Use free or commercial software tools to edit the video (e.g., Adobe Premiere Elements). If you have multiple messages to share, split the video into a series of smaller clips.

- **PowerPoint.** You can use your existing PowerPoint presentations, add animations and transitions between slides, add voice-over and/or music, and save this as your video: no filming needed, although you can, of course, insert videos, and they will work after saving the PowerPoint as a video, whatever device on which the user runs the video.

TABLE 3.9 Examples of how a video (Vlog) can serve your research business

Video/Vlog purpose	Target group
Show your facilities	Grant reviewers
Interview on your new book	Potential readers
Tutorial on a new method	Peers
(Animated) lecture slides	Peers/students
Trailer to promote a conference	Peers/public
MOOC or SPOC	Public/private student group
Short documentary	Public

TABLE 3.10 Etiquette for social networks

> Use a name that represents you well.
>
> Be identifiable; use your real name and a photo.
>
> Separate personal and professional social media.
>
> Don't upload anything that you don't want your peers or boss to see, even to a closed group.
>
> Don't irritate strangers by asking to connect with them without explaining why.
>
> Always remain civil and respectful.
>
> Never post when you are angry, busy, or tired.
>
> Check for spelling mistakes (or worse), especially if you use speech recognition and auto correction.
>
> Respect the conditions for using a specific social media service (e.g., concerning conflicts of interest).

- **Professional films.** Your university is likely to have a studio, equipment, and experts to help you produce a professional video. They'll help you create a "storyboard" to previsualize the story as a series of cartoons with precise instructions for each cartoon on what, who, how, and where to film. They'll film the material and style it into a corporate movie by adding advanced animations and/or professional voice-over or subtitling, for example. And they can help you set up a "short private online course" (SPOC) or a "massive open online course" (MOOC) if that is your goal.

If you film other people, you will need their written consent before you can use and/or distribute the video. Do realize that you are vulnerable and may receive critical reactions and thumbs down. Are you ready for this? And if you have filmed other people, are they ready too? If you made your videos for internal use only, say in a SPOC, you may need them to sign a nondisclosure agreement that forbids external distribution. These are pretty obvious issues, but they are important to consider in good time.

Pitfalls in dealing with media

Imagine that you haven't been very explicit to a journalist that your work is only a first step and that its translation into a practical application will really take years. Soon after publication of the journalist's article, you may be bombarded with phone calls, emails, tweets, and more from people asking for the application (e.g., the treatment of a particular disease). Manage expectations well beforehand.

Although traditional and social media can be beneficial for your research and recognition (see, e.g., the anecdote in Section 3.3), it is important to understand the potential negative consequences as well, in particular, if you break the etiquette rules (see Table 3.10). Even if your etiquette is correct, be prepared for consequences such as heated debates, which may harm your self-confidence if not your public reputation. For example, in 2016, a leading national newspaper published an appeal signed by 180 humanities professors to open

European borders for refugees. Within hours the blog thread exploded with, for example, severe criticism of the "philosophical and historical naivety of these professors in their ivory towers," while the authors had probably hoped for an endorsement of their appeal. The anecdote at the end of this section is another example of how interactions can quickly become heated and dangerous.

The media treat scandals mercilessly. They can damage your reputation and future if they detect or suspect you are not declaring a conflict of interest (collegial, commercial, or other), tweaking or fabricating data, twisting the scientific truth, not properly crediting peers or students, using your power to claim authorship on papers to which you have made no real contribution, or breaching scientific integrity in any way. For example, a professor was found to have fabricated data and had to retract several papers: it was big news, and he got fired. When he later contributed to a blog on Retraction Watch under a false name and posted blogs defending himself, the moderators became suspicious and forced him to declare his real name. The next day he was hot news (again) in many media.

A PROFESSOR'S ANECDOTE

Heated and fired?

My recent blog post on the banking crisis has been read and liked over 10,000 times. However, one person started offending me, and I posted why I thought it was inappropriate. Some more people started harassing me on Twitter, and things went crazy. Those few people accused me of all sorts of things, going well into the range of libel and slander. I'd had enough. The same people went to my personal Facebook page and infiltrated streams of comments. I had to remove these by hand. It was not a good idea, but I told a few of them in private messages to f*** off. Someone then copied this private message to the president of my university. I now feel I am in a rather Kafkaesque situation where I am accused of "absurd behavior against citizens" and I am required to defend myself to "resolve" this matter. What will it take? Will it have implications for my job? Of course I need to take this situation seriously and banish my impulsive behavior in future situations.

TRY THIS!

- Prepare for an interview (e.g., for a local or national newspaper, radio, or TV program, internal or external university newsletter) or for posting on social media (e.g., blog, microblog).

1. **Who.** Step into the shoes of your audience.

   > Educational level(s)
   > Why they want or need to know about your work
   > Recent media attention on your topic
   > Recent letters to editor on your topic
   > Perceptions and emotions with regard to your topic
   > Google the journalist or influencers and communities on social media

2. **What.** Define the information you want to share with your audience.

   > Your main key message (the title of the story, a catchy one-liner)
   > Other key messages (major story elements)

3. **How.** Prepare the story telling.

   > Examples (to make the story more concrete)
   > Anecdotes (to make the story more lively)
   > Metaphors (to clarify by similarity)
   > Quotes (what other people have said and when)
   > Supplementary items, pictures, videos (to visualize your story elements)

4. **Nasty questions.** The questions you don't want to be asked but may have to answer. Label them as questions related to your key messages (k), ones related to your expertise (e), ones on the edge of your expertise (b), or ones you should politely refuse to answer (r).

	Question	Answer
1	What if ...	What I find important is ...
2	If ... then ...	What matters most is ...
3		
4		
5		
6		
7		
8		
9		
10		

- Find the names of science journalists of major (national) newspapers and google them; where do they show up, and can you meet them and get to know them?

- Study Wikipedia rules for adding information about your research to existing Wikipedia pages, or add a Wikipedia page about yourself.

- Google yourself. What shows up? This information is available for future collaborators, employers, journalists, and citizens. What is good, bad, or even ugly? Is your online identity professional, powerful, complete, up-to-date, and visible? Should some of your online accounts be deleted? Should that you improve your style of communication? Is it urgently time to update your web profile?

- If you are a nonuser of social media, try to become a consumer, and analyze who the influential people, communities, or organizations are on Twitter or other media and consider who you could follow. If you are a consumer, try to become an active user (e.g., microblog or blog about your most recent papers; see Table 3.7).

- Use a social media search engine (e.g., Socialmention.com) to find blogs, microblogs, images, and videos that mention your keywords of interest.

3.5

Web Profile

Read what the Nobel Prize Committee writes to those who are invited to nominate a candidate:

There is no need to submit CV, publication list, reprints or other material easily obtained from the internet.

In science, a personal website and a personal email address are your two most important "electronic representations."

Want to know more about me? Check my personal website. Want to contact me? Send me an email.

A professional scientist nowadays needs a professional profile and website more than ever before. Your university probably provides you with a personal page on its website and an email account. These are very useful. But what happens if you move? Do you and your team members actually need a *personal website for life* and an *email address for life*? You're likely to move from an undergraduate course at one university to a graduate course somewhere else, and from there to a first and second postdoc position elsewhere, and maybe on to a fixed-term or permanent position, and maybe work in several more places before you retire. So the answer is, yes, scientists nowadays may want a website and email address for life.

Website content

On your personal webpages you brand and market yourself and your research business. It's the first place where people may learn about your past performance and what you have to offer, whether it is:

- A future employer checking your application letter
- A headhunter matching their job requirements with your web profile
- A reviewer checking the web link from your grant proposal
- A student who searched the web and who is considering working with you
- A scientist in need of partners on a grant application
- A member of a prize committee who might nominate you or one of your team members
- A journalist searching for an expert in your field.

You can provide information on yourself, such as contact information, a description of your research topics, and your CV (Table 3.11a). As soon as you have started your own group, you can scale up and include information on your team and its dynamics. Remember that a (first) impression of your website may feed other people's opinion of your qualities, e.g., the reviewer of your application for a $1 million grant may search for clear proof of your leadership skills and may form an opinion about your personality – all from your website! Does your website profile indeed support your statements about your leadership skills made in the grant application? Can they trust that such a large sum of money will be in good hands? Soft information matters too (e.g., add photos of your team members and social events to demonstrate that you know how much the human aspects of science matter). Provide hard and soft information on your research and yourself as

TABLE 3.11 Examples of what information people are looking for on your personal webpages

3.11a Contact and other basic information

You		
Field of expertise, for specialist and layperson	Job title	Contact info
Short CV	Short CV in PDF	Full address, including room number, phone number, email address

3.11b Quality of research or researcher

Research		
Three to five main scientific achievements	Three to five best publications	All publications
Finished projects	Running projects	Highlighted publications
ORCID Reseacher ID	Link to your Google Scholar account	Altmetrics
Researcher		
Invited lectures	Prizes	Prestigious memberships
Prestigious grants	Grants as PI[a]	Grants as co-PI
Conference organizer	Session chair	Collaborators

[a] *Principal Investigator, meaning you're in charge (and fully responsible).*

3.11c Quality of valorization/impact/engagement/outreach within/outside academia

Valorization		
Patents	Licenses	Spinoff business
Impact		
New consumer products	Improved health outcomes	Policy changes
Engagement		
User involvement	End-user involvement	Citizen science
Press clippings	TV/radio	Social media
Newspaper items	Public lectures	Popular science articles or books

3.11d Quality of mentoring individuals and leading team

Team		
Current people on team	Alumni within/outside academia	Support staff
Visiting scientists	Students/interns	Study with us!
Team seminars	Social events	Retreats
Grants awarded to (former) students/team members	Prizes awarded to (former) students/team members	Work with us! Vacancies and statement on inclusiveness

3.11e Quality of open science

Downloadable		
Articles	Presentations	Videos
Books	Illustrations	Manuals
Software	Data	Metadata

3.11f Transparency on interests

Additional activities		
Journal editor	Commission member	Board member
Ancillary activities		
Consultant	Advisor	Company owner/shareholder
Third parties		
Your collaborations with third parties	Your financial interests in third parties	Your political interests in third parties

researcher (Table 3.11b), your impact on the field (Table 3.11c), your ability to attract, train and mentor students and early-career researchers (Table 3.11d), how you share results (Table 3.11e), and a list of your interests (Table 3.11f).

If allowed, include a proper reference to your university webpage and include the university's logo on your personal website – the prestige and reputation of the university are now stretching out to you. Moreover, this demonstrates that you are a good employee: you wish to serve and promote your university (your alma mater, i.e., "nourishing mother"), too. Upload as much well-ordered information as possible onto your university pages, and add a link to your personal website for life. University webpages are often optimized for search engines: it may be the first hit on the list if someone is searching for ("googling") you.

You can copy any link to external information into your website (e.g., a published paper), and you can add links to your website content (e.g., a video, your public outreach webpage) in a grant proposal; on a poster, a flyer, or your business card; in social media; and so on. Shorten the URL if it has become too long.

- **Short URL.** The Uniform Resource Locator (URL) is a web link. One way of dealing with a long web link is to shorten it. There are free tools into which you can insert your web link and receive a short code in return (e.g., bitly.com or tinyurl.com). This is achieved by using a short redirect on the domain, which links to your webpage with the long URL. Information contained in the URL name is lost, but the short URL is economical for social media, e.g., Twitter, where there is a limit on the number of characters.

- **DOI.** A Digital Object Identifier (DOI) is frequently used by scientific publishers to identify articles. For example, in doi:10.1038/nbt.3240, the prefix 1038 refers to the registrant – Nature Publishing Group – and the suffix refers to article 3240 in the journal Nature Biotechnology (NBT). The DOI links to the URL of the article. The URL can be changed behind the scenes if this be required in the future, whereas the DOI remains the same. DOIs can also refer to a page, table, or figure in an article or to any other digital object. You can only generate DOI codes if you or your university have registered on the DOI system and paid the fee. You could use services provided by CERN to create a DOI yourself (zenodo.org). Tools exist to convert a DOI to other articles IDs such as PubMed's PMID (or from another ID to DOI).

- **DAI.** A Digital Author Identifier (DAI) is increasingly used by scientific publishers, funding agencies, and other organizations to identify researchers. It makes it easy to collect all your publications. It distinguishes you from peers with the same name (e.g., there are many researchers named Li Y). It solves problems with difficult or composite last names (e.g., De Jong may sometimes be written as DeJong and Gonzalo-Morgado as Gonzalo). Get a persistent digital identifier: register at orcid.org to receive your Open Researcher and Contributor ID, and include it when you submit articles, grants, etc. ORCID is nonprofit and supported by many stakeholders. Other IDs, such as ResearcherID and ScopusID, are commercial.

- **QR code.** The Quick Response Code (QR Code) is a two-dimensional barcode for digital information of limited size, e.g., up to 7,000 characters. Users can scan the QR Code with their smartphone, which may then show other relevant information stored in the code, or if it encodes a URL, it will automatically open a website, start a video, and so on. You can add the image of a QR Code to any document. For example, add it to your grant proposal, and panel members and reviewers may scan it to see a short video in which you show them your facilities or briefly pitch your project idea. Add it to your conference poster, and conference participants can easily open your website with further information on your research. Scanning a QR image is much easier than typing a long URL into your smartphone's browser. There are several free and paid websites and apps to generate QR Codes (e.g., Google). Smartphones have built-in apps for scanning QR Codes, or they can be freely downloaded.

Email

Email has become a dominant medium for communication in science, and on a daily basis you (and your team members) send and receive many emails. Your email address is also shown on published scientific articles (readers can contact you with queries), on your scientific posters, on the participant lists of conferences, on your business card, on the university's website and your personal page, and so on. Additional information may be attached to the email or via links to websites. Consequently, your email boxes serve as an important electronic archive for your research business, and there is little need to delete emails nowadays. Set an automatic forward from your university email account to your personal account. *Note:* Your university may well have a system for backing up sent and received email and for recovering deleted emails or folders; other providers may not provide this backup, and you must arrange it yourself. But once you move, your university archive may be closed and deleted.

Sensitive personal issues should not be discussed by email; rather, use the phone or go to speak directly to your team member, director, or collaborator. You don't want to risk the recipient misunderstanding your message or – worse – that he or she shares it with other people or – even worst – uses it in court. Opt for direct face-to-face or ear-to-ear contact, in which misunderstandings can be detected and ironed out immediately. Much of human communication is not in the words themselves but in the tone, timbre, and speed of voice and all the other body language signals that you miss in a written email but should observe to become a (more) effective team leader. In some cases, you may, however, prefer to use email so that you have time to think about and fine-tune your message. Let it rest for a few hours, and read your draft email again as if you are the recipient, and only then send it. And be careful with "Reply to all," especially when it regards a heated debate.

Busy people such as directors or deans may receive hundreds of emails per day, and if they don't recognize your name, they may not even open the email. Therefore, it is extremely important to choose the right subject line for your email. It should be informative and catchy enough to raise curiosity to open your email first. The first two or three sentences should be strong and convince the reader to read more. Emails to busy people should be simple and short with a transparent layout. Actually, we're all pretty busy, so why not discuss expectations on internal email behavior (Table 3.12)?

Your emails can really look like webpages: a nice layout, hyperlinks to your and other people's websites, with photos, pictures, and icons for social media. This type of enhanced email is especially appropriate for newsletters or event announcements. You can spread the word about your research group or about the conference you're organizing in a catchy format to a dedicated group of people. Good HTML email templates that work for many types of displays are freely available. It is best to also include a normal text version in your newsletter for those who can't receive or view the HTML version (many email servers do this automatically for you) and to offer them a link to the website where they can open and view the content in their internet browser (you have to add this link to the HTML version). Special mailing list software is needed if you want to fully or partially automate sending emails to a list of receivers (e.g., MailChimp). This software uses a "reflector" email address: you send your email to this reflector email address to distribute it to all the addresses on the corresponding emailing list – the recipients will only see the reflector email address and not the addresses of the other recipients. For a newsletter, users may subscribe or unsubscribe through a web-based interface or by sending an email to a special email address.

There are two ways to protect the privacy of your email communication if necessary.

- **Add a disclaimer.** Before you know it, you've selected the wrong address from a dropdown list. At a minimum, you can attach a (link to a) disclaimer to important emails explicitly instructing an unintended receiver to ignore the email (see Table 3.13). Disclaimers have a legal status.

TABLE 3.12 Agree with your students, team members, and support staff on keeping internal emailings manageable: examples of what can be agreed on for each of the five components of an email

Sender/receiver	Avoid massive emailing. Send fewer emails. Use "cc" and "Reply to all" sparsely. Don't email information that should be saved for meetings.
Date/time	Reply within two working days. In the exceptional case that one needs a quicker reply, begin the subject line with "URGENT."
Subject line	Use strong keywords. A receiver can then decide to read now or to retrieve it later when the time is right. Instruct senders to use, e.g., "report-week3."
Email body	Make it easy to read. Describe the main purpose of the email in the first three lines. Structure your message. Use headings, short paragraphs, blank lines, bullets, bold, underline, and/or italic. End with a short informative signature.
Email body (reply)	Make it easy to read. Edit a reply email such that it can be read top down; e.g., first their question, thereafter your answer. Remove signatures and other irrelevant context from the original email.
Attachments	Use informative file names. A receiver can save such files without renaming. Instruct senders to use, e.g., "report3-their-name.doc." Don't attach a logo or signature.

- **Seal your email.** At the push of your "Send" button, your text is transferred from your computer via a set of computer nodes to the inbox of the receiver – just as with ordinary mail – and unless carefully sealed, all information can be read like the back

TABLE 3.13 Examples of disclaimers to use in your email: from basic to more extensive (extensions in italic)

The contents of this message are confidential and only intended for the eyes of the addressee(s). Others than the addressee(s) are not allowed to use this message, to make it public, or to distribute or multiply this message in any way. The University of XXX cannot be held responsible for incomplete reception or delay of this transferred message.
The information transmitted, including attachments, is intended only for the person(s) or entity to which it is addressed and may contain confidential and/or privileged material. Any review, transmission, dissemination, or other use of, or taking any action in reliance on, this information by persons or entities other than the intended recipient is prohibited. *If you received this in error, please contact the sender and destroy any copies of this information.*
The information contained in this communication is confidential and may be legally privileged. It is intended solely for the use of the individual or entity to whom it is addressed and others authorized to receive it. If you are not the intended recipient, you are hereby notified that any disclosure, copying, distribution, or taking any action in reliance of the contents of this information is strictly prohibited and may be unlawful. If you have received this message in error, please contact the sender immediately by return email. University of XXX is neither liable for the proper or complete transmission of the information contained in this communication nor for any delay in its receipt. *University of XXX has taken every reasonable precaution to ensure that any attachment to this email has been swept for viruses. However, we cannot accept liability for any damage sustained as a result of software viruses and would advise that you carry out your own virus checks before opening any attachment. This email is meant to communicate company-related materials only. Non-business-related opinions expressed by the author of this email are solely his/her own. University of XXX will not be liable for such opinions expressed in this email.*

of a traditional postcard. We all know that companies such as Google and government agencies such as the CIA can monitor your emails and perhaps even read your texts. So, too, may other parties. You can encode your information by encryption, which requires the sender and receiver to collaborate on sending public and private keys.

Use existing web platforms?

Several commercial parties now offer platforms for sharing professional/research profiles, for example, LinkedIn, Wordpress, ResearchGate, Academia.edu, and Mendeley.com. You can get your publications online easily. It's probably wise to use one or more of these academic platforms because it increases the chance of being found, cited, invited, scouted, and so on. However, open science conflicts with such commerce and the associated lack of transparency and accountability on associated metrics. Commercial parties need to make a profit at some point, just like the companies that see academic publishing as a highly profitable market. Moreover, companies can change their policies and strategies, can be sold, or can even go bankrupt: academia should not become dependent on them. In contrast, universities are the most stable of institutions, some are centuries old. Your university – or a consortium of universities – could potentially provide you with a stable system with a lifetime website and email address.

Webpages for support staff?

Should key support staff provide personal pages of their own and at a level and style similar to those of scientists? It might help us understand how serious we all are about making science our business. So the answer is, "yes they should." Their intranet webpage can show clear evidence of their quality of work. Also, the support department's intranet should show concrete information about what services are offered, what impact on research and researchers these services have

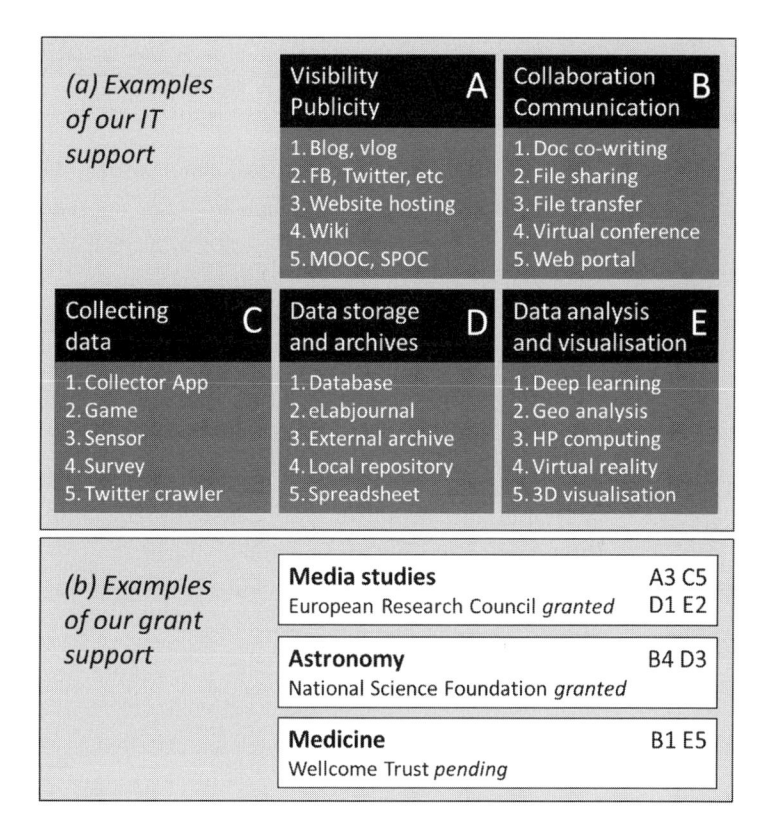

(a) Examples of our IT support	Visibility Publicity A	Collaboration Communication B
	1. Blog, vlog 2. FB, Twitter, etc 3. Website hosting 4. Wiki 5. MOOC, SPOC	1. Doc co-writing 2. File sharing 3. File transfer 4. Virtual conference 5. Web portal
Collecting data C	Data storage and archives D	Data analysis and visualisation E
1. Collector App 2. Game 3. Sensor 4. Survey 5. Twitter crawler	1. Database 2. eLabjournal 3. External archive 4. Local repository 5. Spreadsheet	1. Deep learning 2. Geo analysis 3. HP computing 4. Virtual reality 5. 3D visualisation

(b) Examples of our grant support	**Media studies** European Research Council *granted*	A3 C5 D1 E2
	Astronomy National Science Foundation *granted*	B4 D3
	Medicine Wellcome Trust *pending*	B1 E5

FIGURE 3.6 Pre- and postaward services provided by a center for information technology. "We have helped dozens of researchers with writing grant applications and running awarded projects."

had previously, and how confident the researchers were (Figure 3.6).

A PROFESSOR'S ANECDOTE

Sh*t happens

I was expecting emails with information needed to finalize a grant proposal with multiple partners. I was also organizing a conference and expected emails from the speakers with their abstracts. This is the usual type of stuff we're often involved in. Nothing special.

My university used email software produced and delivered by one of the global players in the field. This was very reassuring. From time to time, upgrades were needed, and sometimes they were major. Well, we're one of many clients, so why should that be a problem?

And so, on the announced date, the system was upgraded. And then the trouble started. At first, people logging on were connected to other people's email boxes. Then the incoming emails were bounced. And finally, the whole system broke down. Not just for an hour, nor for a day, but for an entire week! Without any information about when or whether it would be fixed.

Without email, the university was lost; it simply couldn't function anymore. The crisis had begun, although not for all of us. Some – including me – were having their emails forwarded automatically to a second, independent account, and luckily, the forwarding still worked even after the system crashed. Hurray. And so, although far from ideal, this backup system helped me to access my email and survive while other colleagues continued to suffer.

Such crises happen, but are we ready for them? Do we rely too heavily on companies such as Google or Microsoft, who can drastically change their terms and conditions – or even fall apart? Can we resist distributed denial of service (DDoS) attacks on our institution? What else should we be prepared for in these high-tech social networking days? Attacks such as in June 2017 on Rotterdam harbor (NL) and the National Health Service NHS (UK) make the threat very concrete.

TRY THIS!

Website

- The Uniform Resource Locator (URL) is the address of your website. It should be informative about the content of the website and at the same time be short to type and simple to memorize. Create a unique selling URL for your *website for life*, and check whether it is still free (use a search engine). It consists of three parts: www.name.extension. For the name part, you may want to use your own name, e.g., "smithlab" or "anthonysmith," or a short text pitching your niche, e.g., "lawcasestudies" or "lawcasesportal" (although an addition such as "portal" may suggest that it's the *entrance* site to this whole field). Extensions can be "edu," "org," "com," "info," or the abbreviation for your country (e.g., "eu" or "uk").

- Develop a homepage and subpages structure.

- Write quality content for each page.

- Create a brief title for each page (<20 words).

- Write an accurate summary for each page (<30 words).

- Think of the best search terms to add to your webpage (search engine optimization [SEO]).

- Decide to ask experienced support staff about examples of previous work, timeframes, and costs. Or decide to do it yourself and search for "free website hosting" and "free website tools." For example, Wordpress is a content management system (CMS) with ready-to-use templates for academic websites that is free to use under a CC-BY license.

- Check the result a couple of weeks after the website has gone online: does your site pop up well on search engine lists for relevant queries? Check several different search engines on different computers (otherwise, your search history influences what pops up).

- Check the other statistics after some time: (a) how many people visit your site, from which links do they access it, where do they land, and how long do they stay?; (b) how many external sites link to yours?; and (c) what queries drive people to your site? Your provider and search engines (e.g., Google Analytics) can provide these statistics.
- Universities may have licenses on advanced reporting systems (e.g., PURE) that can produce all kinds of reports (e.g., list of publications) and place them on your webpage on the university's website.

Email

- Create your email address for life, perhaps one derived from the URL of your website for life. The provider hosting your personal website will usually also offer one or more email addresses. See Table 3.14 (the email address also consists of three parts: localname@name.extension).
- Develop a short and catchy signature for your emails. Include essential web links and information on your recent successes.

TABLE 3.14 Options for your email address

info@smithlab.org	May end up in spam folders
anthony@smithlab.org	More specific and personal
courses@smithlab.org	Use more than one local name
smithlab@gmail.com	If you want to use Google
amsmith@gmail.com	Using initials and surname
amsmith2019@gmail.com	Add year to reduce spam

- Do you need to include a disclaimer (or link to a disclaimer) in your email? Prepare one, and discuss it with your legal affairs officer.

- "Sensitive personal issues should not be discussed by email; rather, use the phone or go to speak directly to your team member, director, or collaborator." Check your recent email history. How often did you email while using the phone or speaking fact-to-face would have been more appropriate?

Wikipedia

- Read the "professor test" criteria in Wikipedia's guidelines for academic subjects or biographic articles to see if you can be included on Wikipedia: https://en.wikipedia.org/wiki/Wikipedia:Notability_%28academics%29.

- Read Wikipedia's conflict-of-interest policy: when to refrain from contributing/editing Wikipedia content: http://en.wikipedia.org/wiki/Conflict-of-interest_editing_on_Wikipedia.

- Are your research field, your research (articles), and/or your personal information covered in Wikipedia? Should you consider contributing yourself?

Epilogue

Research can move the frontiers of knowledge if – and only if – the new knowledge is properly transferred to users who want or need it. This is what academia is all about.

The word "academia" derives from Greek Ἀκαδήμεια. This was a domain dedicated to the Athenian hero Akademos at a site with a καδημία, a sanctuary dedicated to Athena, the goddess of wisdom. Here Plato founded his school of philosophy around 387 BC. The word "philosophy" derives from Greek φιλοσοφία, originally meaning "love of wisdom," and the word "school" derives from Greek σχολή, originally meaning "leisure." Plato's disciples – wealthy men who had leisure – came of their own free will to the school of philosophy to gather knowledge and think critically about it.

Many centuries later you run your own "school of philosophy" in academia. You train your Bachelor's and Master's degree students, PhD candidates, and postdocs. This costs you time that, in return, leverages your research.

Every year there is an enormous influx of new *students* and a major outflow of an educated workforce. Helping all these (early-career) people make the right career choices and fully develop their talents is crucial to them and to society – and for your research too. Some students show a talent for science, and such students will make your heart beat faster. A very small fraction of motivated students (<1 percent) is invited to extend their career in science, and they may work with you for some time: a Bachelor's degree student becomes a research Master's degree student and later a PhD student. Very few go on to become full professors. The students and postdocs are your workforce, and they help you boost your research output until they bid you farewell for a new position – in academia or elsewhere. If you are the team leader, then you have a critical responsibility for educating your *team members* to prepare them not only for their role on your team but also for their next career step. The career pyramid in science is steep; many postdocs and assistant and associate professors on tenure-track positions will have to continue their careers outside academia. The university is a "learning and teaching" environment, and saying farewell to most people is just part of the business model. This is an unfortunate knowledge drain from the researcher's viewpoint, but this is why the government is able to dedicate financial resources specifically to you and your peers in academia.

Your educational responsibilities go beyond students and team members. As soon as the frontiers of knowledge have moved – which is what you are aiming to achieve – you also have to disseminate this new knowledge to those who have already left: the *alumni* of your and other universities. Lifelong learning is increasingly important in today's knowledge-based society.

You have a responsibility not only to those who have left but also to those who are considering joining: *high school students*. Only with your help can they make a rational choice to study at university. Good for them! Good for society! The benefit to you? A better fit between you and your students and a better resource from which to recruit your future students and team members.

How about the nonexperts, the *lay audience*, the men and women on the street – do you need to educate them? They often lack the basic background and may not understand the keywords used in your research. Still, the general public should be able to get a basic understanding of what your research is about. After all, they're paying for your research with their taxes. As citizens, they also vote, and via their elected politicians, they will decide about society's grand challenges, their translation into research agendas, and the allocation of budgets for university research and business. Imagine how this might work out if your type of research is not on the agenda.

You have a lot of freedom in academia, so be entrepreneurial: show initiative, and develop your strategy to make the most of your "business." But there's only so much you can know and handle yourself. Luckily, *support staff* can offer crucial expertise and extra hands. Therefore, a first step in your strategy should be to establish and nourish productive relationships with them. Turn your "business" into a joint venture!

Further Reading

Books

Cain S (2012). *Quiet: The Power of Introverts in a World that Can't Stop Talking.* New York, NY: Crown Publishing Group.

Covey SR (1989). *The Seven Habits of Highly Effective People.* New York, NY: Simon & Schuster.

Covey SR (2004). *The 8th Habit: From Effectiveness to Greatness.* New York, NY: Free Press.

Eve MP (2014). *Open Access and the Humanities.* Cambridge: Cambridge University Press. https://doi.org/10.1017/CBO9781316161012.

Fisher R, Ury WL, Patton B (1991). *Getting to YES: Negotiating an Agreement without Giving In* (2nd edn). New York, NY: Penguin Group.

Heath C, Heath D (2007). *Made to Stick: Why Some Ideas Survive and Others Die.* New York, NY: Random House.

Jansen RC (2011). *Developing a Talent for Science.* Cambridge: Cambridge University Press. https://doi.org/10.1017/CBO9780511667053.

Jansen RC (2013). *Funding Your Career in Science: From Research Idea to Personal Grant.* Cambridge: Cambridge

University Press. https://doi.org/10.1017/CBO9781139626965.

Kolb DA (1983). *Experiential Learning: Experience as the Source of Learning and Development*. Upper Saddle River, NJ: FT Press.

Maister DH, Green CH, Galford RM (2000). *The Trusted Advisor*. New York, NY: Free Press, Simon & Schuster.

Riso DR, Hudson R (2017). *Personality Types*. Old Saybrook, CT: Tantor Media.

Suber P (2012). *Open Access*. (MIT Press Essential Knowledge Series). Cambridge, MA: MIT Press. Available at http://legacy.earlham.edu/~peters/fos/overview.htm.

Ury WL (2007). *Getting Past NO: Negotiating with Difficult People* (2nd edn). New York, NY: Penguin Group.

West MA (2012). *Effective Teamwork: Practical Lessons from Organizational Research* (3rd edn). Leicester: BPS Blackwell.

Yuki G (2013). *Leadership in Organisations* (8th edn). London: Pearson Education.

PLoS journals

Articles, commentaries, and perspectives published by the Public Library of Science (PLoS). For example:

McKiernan EC (2017). Imagining the "open" university: Sharing scholarship to improve research and education. *PLoS Biol* 15(10):e1002614. https://doi.org/10.1371/journal.pbio.1002614.

Moher D, Naudet F, Cristea IA, et al. (2018). Assessing scientists for hiring, promotion, and tenure. *PLoS Biol* 16(3): e2004089. https://doi.org/10.1371/journal.pbio.2004089. See also the "Ten Simple Rules" collection published by PLoS Computational Biology:

Bik HM, Dove ADM, Goldstein MC, et al. (2015). Ten simple rules for effective online outreach. *PLoS Comput Biol* 11(4): e1003906. https://doi:10.1371/journal.pcbi.1003906.

Hart EM, Barmby P, LeBauer D, et al. (2016). Ten simple rules for digital data storage. *PLoS Comput Biol* 12(10):e1005097. https://doi:10.1371/journal.pcbip.1005097.

Martin JL (2014). Ten simple rules to achieve conference speaker gender balance. *PLoS Comput Biol* 10(11):e1003903. https://doi:10.1371/journal.pcbi.1003903.

Michener WK (2015). Ten simple rules for creating a good data management plan. *PLoS Comput Biol* 11(10):e1004525. https://doi:10.1371/journal.pcbi.1004525.

Shaw DM, Erren TC (2015). Ten simple rules for protecting research integrity. *PLoS Comput Biol* 11(10):e1004388. https://doi:10.1371/journal.pcbi.1004388.

Journals *Nature* and *Science*

Articles and commentaries published by the journals *Science* and *Nature* (some on their NatureJobs pages). Check for more recently published articles and commentaries. Here is a snapshot in this collection:

Bosch G (2018). Train PhD students to be thinkers not just specialists. *Nature* 554:277. https://doi:10.1038/d41586-018-01853-1.

Collins FS, Tabak LA (2014). Policy: NIH plans to enhance reproducibility. *Nature* 505:612–13. https://doi:10.1038/505612a.

Editorial (2018). A code of ethics to get scientists talking. *Nature* 555:5. https://doi:10.1038/d41586-018-02516-x.

Editorial (2018). Checklists work to improve science. *Nature* 556:273–74. https://doi:10.1038/d41586-018-04590-7.

Flier F (2017). Faculty promotion must assess reproducibility. *Nature* 549:133. https://doi:10.1038/549133a.

Merton DK (1968). The Matthew effect in science. *Science* 159 (3810):56–63. https://doi:10.1126/science.159.3810.56.

Open Science Collaboration (2015). Estimating the reproducibility of psychological science. *Science* 349(6251). https://doi:10.1126/science.aac4716.

Powell K (2015). Interviews: Big ideas for better science. *Nature* 528: 589–90. https://doi:10.1038/nj7583-589a.

Price M (2018). Proposal to rescue postdocs from limbo draws darts. *Science* 360(6386):253. https://doi:10.1126/science.360.6386.253.

Special issue (2015). Challenges in irreproducible research. *Nature* (growing online list of articles). www.nature.com/news/reproducibility-1.17552.

Special issue (2015). Interdisciplinarity. *Nature*. www.nature.com/news/interdisciplinarity-1.18295

Stephan P, Veugeler R, Wang J (2017). Reviewers are blinkered by bibliometrics. *Nature* 544: 411–12. https://doi:10.1038/544411a.

Sundin J, Jutfelt F (2018). Keeping science honest. *Science* 359 (6383):1443. https://doi:10.1126/science.aat3473.

Urry M (2015). Science and gender: Scientists must work harder on equality. *Nature* 528:471–73. https://doi:10.1038/528471a.

Woolston C (2015). Conference tweeting rule frustrates ecologists: Complaints ensued when attendees at an ecology meeting were asked to get permission before live-tweeting. *Nature* 524:391. https://doi:10.1038/524391f.

Woolston C (2018). Science PhDs lead to enjoyable jobs. *Nature* 555:277. https://doi:10.1038/d41586-018-02696-6.

Other scientific journals

Bol Th, De Vaan M, Van De Rijt A (2018). The Matthew effect in science funding. *PNAS*, published ahead of print April 23. https://doi.org/10.1073/pnas.1719557115.

Eve MP, Inglis K, Prosser D, Speicher L, Stone G (2017). Cost estimates of an open access mandate for monographs in the UK's third Research Excellence Framework. *Insights* 30 (3):89–102. https://doi:10.1629/uksg.392.

Gaucher D, Friesen J, Kay AC (2011). Evidence that gendered wording in job advertisements exists and sustains gender inequality. *Journal of Personality and Social Psychology* 101 (1): 109–28.

Munafo MR, Nosek BA, Bishop DVM, et al. (2017). A manifesto for reproducible science. *Nature Human Behavior* 1. https://doi:10.1038/s41562-016-0021.

Prins P, De Ligt J, Tarasov A, et al. (2015). Toward effective software solutions for big biology. *Nature Biotechnology* 33:686–87. https://doi:10.1038/nbt.3240.

Redeker M, De Vries RE, Rouckhout D, Vermeren P, De Fruyt F (2014). Integrating leadership: The leadership circumplex. *European Journal of Work and Organizational Psychology* 23: 435–55.

Scheffer M (2014). The forgotten half of scientific thinking. *PNAS* 111:6119. https://doi:10.1073/pnas.1404649111.

Trelles O, Prins P, Snir M, Jansen RC (2011). Big data, but are we ready? *Nature Reviews Genetics* 12(3):224. https://doi:10.1038/nrg2857-c1.

Tuckman BW (1965). Developmental sequence in small groups. *Psychological Bulletin* 63(6):384–99. http://dx.doi.org/10.1037/h0022100.

Wilkinson MD, Dumontier M, Aalbersberg IJ, et al. (2016). The FAIR guiding principles for scientific data management and stewardship. *Scientific Data* 3:160018. https://doi:10.1038/sdata.2016.18.

Commentaries in *The Scientist* and *Scientific American*

Check for their daily or weekly articles and commentaries. Here is a snapshot:

Akst J (2015). #IceBucketChallenge payoff: A year after the viral ALS awareness campaign raised millions, researchers say they have something to show for the influx of funds. *The Scientist*, August 20.

Grant B (2014). The working vacation: Sabbaticals are one of the perks of the academic life. They may seem daunting to implement, but the time away could prove invaluable to your career. *The Scientist*, April 1.

Grens K (2015). Self-correction: What to do when you realize your publication is fatally flawed. *The Scientist*, December 1.

Kwon D (2017). A turbulent year in the publishing world. *The Scientist*, December 16.

Lee DN (2014). Scientists draw on personal experience to guide their curiosity. *Scientific American*, October 1.

Phillips KW (2014). How diversity makes us smarter. *Scientific American*, October 1.

Pritsker M (2013). Opinion: Video saved the scientific publication. How visual materials and methods can save scientists time and money. *The Scientist*, November 11.

Thomas JR (2015). Scientific misconduct: Red flags. Warning signs that scandal might be brewing in your lab. *The Scientist*, December 1.

Yandell K (2014). Keeping up with IP. It's never too early to start thinking about intellectual property rights – even for biologists doing basic research. *The Scientist*, September 1.

Vence T (2015). Know your PIO. Scientists and public information officers share several common goals. Here's how to collaborate effectively. *The Scientist*, January 1.

Blogs

Scientists blog on all kind of topics. Below is a snapshot of blogs on topics related to this book. Check social media for new blogs.

Albers E (2016). There is no open science without the use of open standards and free software. June 2. http://blog.3rik.cc/2016/06/there-is-no-open-science-without-the-use-of-open-standards-and-free-software/.

Anonymous academic (2015). Academics, you need to be managed. It's time to accept that. August 21. www.theguardian.com/higher-education-network/2015/aug/21/academics-you-need-to-be-managed-its-time-to-accept-that.

Anonymous academic (2016). Academia is now incompatible with family life, thanks to casual contracts. December 2. www.theguardian.com/higher-education-network/2016/dec/02/short-term-contracts-university-academia-family.

Bodewits K, Gramlich F, Giltner D (2018). The autopilot postdoc. Falling into a postdoc after a PhD is a waste of excellent credentials. March 26. http://blogs.nature.com/naturejobs/2018/03/26/the-autopilot-postdoc/.

FURTHER READING

Cameron D (2013). Solving big-data bottleneck. Scientists team with business innovators to tackle research hurdles. February 7. https://hms.harvard.edu/news/solving-big-data-bottleneck-2-7-13.

Donald A (2016). Writing reference letters. Do you want to be described as hard working? December 2. www.timeshighereducation.com/blog/gendered-adjectives-do-you-want-be-described-hard-working.

Dunleavy P (2015). How to write a blogpost from your journal article. May 17. https://medium.com/advice-and-help-in-authoring-a-phd-or-non-fiction/how-to-write-a-blogpost-from-your-journal-article-6511a3837caa#.osq5tvb8r.

Elson M (2016). Retaining copyright for figures in academic publications to allow easy citation and reuse. November 11. https://medium.com/@malte.elson/retaining-copyright-for-figures-in-academic-publications-to-allow-easy-citation-and-reuse-77c6e2b511fe#.o0odm3ek5.

Macfarlane B (2014). No place for introverts in the academy? In today's brave new world of university learning, students aren't allowed to be shy. September 25. www.timeshighereducation.com/features/no-place-for-introverts-in-the-academy/2015836.article and www.timeshighereducation.com/features/the-story-of-a-shy-academic.

Martin J (2016). Conference gender speaker balance. Show me the policy (part 2). December 9. https://cubistcrystal.wordpress.com/2016/12/09/show-me-the-policy-part-2/.

Masuzzo P, Martens L (2017). Do you speak open science: Resources and tips to learn the language. January 3. *Peer J Preprints* 5:e2689v1. https://doi.org/10.7287/peerj.preprints.2689v1.

Miedema F (2018). Setting the agenda: "Who are we answering to?" BMJ Blogs, January 24. http://blogs.bmj.com/open

science/2018/01/24/setting-the-agenda-who-are-we-answering-to/.

Retraction Watch (2017). "I placed too much faith in underpowered studies" Nobel Prize winner admits mistakes. February 20. http://retractionwatch.com/2017/02/20/placed-much-faith-underpowered-studies-nobel-prize-winner-admits-mistakes/.

Shmerling RH (2016). Medical news a case for skepticism. April 22. www.health.harvard.edu/blog/medical-news-a-case-for-skepticism-201604229481.

Valatine HA (2016). How to fix the many hurdles that stand in female scientists' way. Women face discrimination of many kinds. We need a culture change. December 1. www.scientificamerican.com/article/how-to-fix-the-many-hurdles-that-stand-in-female-scientists-rsquo-way/.

Guides, platforms, toolkits, and more

Websites last accessed April 22, 2018:

10 Principles of citizen science, a collection of citizen science guidelines and publications: ecsa.citizen-science.net/documents.

Citizen science projects: www.socientize.eu (see also white paper).

Collective Labour Agreement Dutch Universities: www.labouragreementuniversities.nl (see also the full text of the agreement).

Crowdfunding platform to support open access: www.knowledgeunlatched.org/.

DORA (Declaration on Research Assessment): https://sfdora.org/.

European Open Science Cloud: http://ec.europa.eu/research/openscience/eosc.

How open is it? A guide for evaluating the openness of journals, SPARC, the Scholarly Publishing and Academic Resources Coalition: https://sparcopen.org/our-work/howopenisit/.

Knowledge exchange and impact: https://info.lse.ac.uk/staff/services/knowledge-exchange-and-impact/KEI-Toolkit.

Aalbersberg IJ, et al. (2018, February 15). *Making Science Transparent by Default; Introducing the TOP Statement.* http://doi.org/10.17605/OSF.IO/SM78T.

Working Group of the Steering Group of Human Resources Management in the European Research Area (2015). Open, Transparent and Merit-Based Recruitment of Researchers: Checklist and Report: https://cdn5.euraxess.org/sites/default/files/policy_library/otm-r-checklist.pdf and https://cdn1.euraxess.org/sites/default/files/policy_library/otm-r-finaldoc0.pdf.

Peer Reviewers' Openness Initiative: https://opennessinitiative.org/.

Public Engagement: www.publicengagement.ac.uk.

Acknowledgments

I offered earlier versions of this book to many people. They provided me with their valuable feedback on the particular sections of their expertise or in some cases on the entire draft.

Special thanks to the following scientists (PhD candidates, postdocs, assistant professors, associate professors, full professors, research directors, faculty deans, a university rector, and alumni): Casper Albers, Anouk Baars, Jim Coyne, Vera de Bel, Lude Franke, Erik Frijlink, Harry Garretsen, Andreas Herrmann, Marian Joëls, Merel Keijzer, Frank Miedema, Ingrid Molema, Ellen Nollen, Alison Perry, Theunis Piersma, Mladen Popović, Pjotr Prins, Diederik Roest, Ody Sibon, Elmer Sterken, Gerben van der Vegt, Connie van Ravenswaaij, Yvonne Verkuijl, Gerry Wakker, and Martijn Wieling.

Special thanks also to the following support staff (consultants in career counseling, research funding, information technology support, talent development; trainers in grant proposal writing, talent development; policy advisors in human resources, legal affairs, open access, research funding, research strategy, talent development; directors of alumni relationships, communication and marketing, education and research strategy, finance and control, human resources, university library): Geert-Jan Arends, Holger Bakker, Aize Bouma, Heidi Disler, Margot Edens, Jan Feringa, Hans Gankema, Ineke Ganzeveld, Eric Hoogma, Esther Hoorn, Gerald Lier, Tienke

Koning, Peter Meister-Broekema, Bregje Mollee, Marjolein Nieboer, Frank Nienhuis, Klazien Offens, Anke Schuster-Koster, Marion Stolp, Dicky Tamminga, Grytsje van der Meer, Peter van Laarhoven, Marijke Verheij, Yvonne Verkuijl, and Liesbeth Volbeda.

And finally, very special thanks to Katrina Halliday (commisioning editor at Cambridge University Press), Eva Jordans (leadership specialist), Jackie Senior (science editor, editor of this book), and Henny van Zanten (married to author).

Index

academia, Greek Ἀκαδήμεια, *202*
administrative assistant, *58*
agreement
  confidentiality, *111*, *126*
  consortium, *48*, *112*, *117*, *170*
  grant, *112*, *123*, *170*
  labor, *109*, *110*, *117*, *122*
  memorandum of understanding, *113*
  non-disclosure, *111*, *180*
  student, *123*
  visitor, *123*
Altmetric, *176*
alumni, engaged, *159*
alumni officer, *160*
appraisal interview, *81*
articles
  checking the licensing model, *144*
  version control, *147*
arXiv.org, *146*
award system, skewed, *162*
AWIS, American Association for Women in Science, *71*

branding
  personal, *138*
  the university's, *169*
buddy, *60*, *86*
BY (made by me) license option, *142*

candidate
  benchmarking a, *30*
  with alternative career, *33*
  with career break, *33*
career
  action plan, *90*
  courses, *92*, *95*
  fair, *16*, *17*, *22*,
  in/outside academia, *63*,
  progression of team member, *25*
  progression of yourself, *63*
  role of publications, *138*
CC license, Creative Commons, *144*, *153*
citizens
  contributing to project, *154*
  giving unsolicited opinions, *158*
  in review panel, *158*
  in steering committee, *158*
  managing their expectations well, *163*
  strategy to get them involved, *160*
CMS, content management system, *199*
code of conduct for good research practices, *40*
  additional or ancillary activities, *48*, *53*
  being honest and accountable, *47*, *52*
  being independent and critical, *52*
  financial interests, *48*, *49*
  good practices as team leader, *61*
  third-party interests, *48*

code of conduct for good (cont.)
  unethical use of research funding, *102*
  working with other teams, *60*
  your university's, *52*
code of ethics. *See* code of conduct for good research practices
commercial applicability, *125*
communication officer, *169*, *175*
communication, the who, what and how of, *173*
confidential advisor, *89*
confidentiality
  being clear about, *119*
  in agreements, *109*
  of data and documents, *115*, *120*
consortium
  of universities, *196*
  with peers, *117*
  with third parties, *48*, *117*
constructive controversy, *59*
consultancy, what to charge, *105*
copyright, *114*, *117*
  license options BY, SA, NC, ND, *142*
  when it holds, *119*
creative work, *117*, *125*
cross-disciplinary research
  experiences, *67*
  financial barriers for collaboration, *104*
  fostering, *96*
crowdfunding, *154*
  citizens organizing, *163*
  donations by alumni, *160*
  platforms, *156*
crowdsourcing, *154*
  platforms, *158*
culture
  influence on research questions, *44*
  people can talk about personal problems, *89*
  working together with support staff, *70*
  your group's, *35*

DAI, Digital Author Identifier, *191*
data management
  center, *149*, *153*
  plan, *49*, *150*
data management officer, *149*
Dataverse, data repository, *149*
DDoS, distributed denial of service, *198*
depreciation time, *101*, *103*, *107*
disability, *34*, *39*
diversity, *36*, *96*
DOI, Digital Object Identifier, *149*, *190*

email
  adding a disclaimer, *195*
  address for life, *185*, *200*
  keeping manageable, *194*
  protecting privacy, *193*
  system breaking down, *198*
EPO, European Patent Office, *130*
Epoline, *134*
ERC, European Research Council, *63*
Espacenet, *130*
EU Erasmus Program, *17*
EU HR logo of excellence, *82*
evaluation interview, *84*
excellence in research, *42*
exclusive rights
  patent, *126*
  publishers, *117*
expenditures, *100*

Facebook
  connecting with citizens, *162*
  groups, *139*
  job post, *22*
FAIR data principles, *150*
family, starting a, *35*
financial officer, *6*, *97*, *107*
financial processes, mandatory
  before starting project, *100*
  while running and closing project, *102*
funding, personal plan, *29*, *30*
funding officer, *97*

INDEX

GDPR, General Data Protection Regulations, *81*, *96*, *149*, *176*
gender
  bias, *34*, *38*
  equality, *37*
GitHub, software repository, *149*
go/no-go job contract decision, *89*
Google
  AdWords, advertising job, *15*, *22*
  Analytics, *200*
  check for patents, *134*
  curiosity-driven activities, *64*
  Scholar, *130*
  yourself, *183*
grant application, writing a, *5*, *64*,
GRC, Gordon Research Conferences, *123*
group
  ideal composition, *19*
  is not a team, *57*
guest researcher, *91*

habits and behavior, reprogramming, *44*
headhunting
  becoming your own headhunter, *15*
  software, *15*
health coach, *60*
high potentials, *15*, *19*
human resources officer, *6*, *22*, *80*, *89*
human resources policies
  advertisement policies, *83*
  end of project policies, *90*
  evaluation policies, *95*
  performance policies, *84*
  recruitment policies, *82*, *94*
  selection committee policies, *83*
  training and advancement policies, *90*
human resources processes, mandatory
  during closing project, *92*
  during running project, *86*
  for recruiting PhD or postdoc, *85*

idea, new, *124*
  keeping confidential, *110*
  obligation to report, *125*
  ownership, *117*
illness, *101*
inclusion, *36*, *96*
information management plan, *149*
information technology expert, *149*
infringement
  of contract by you, *125*
  of patent by others, *131*
intellectual property
  examples of, *114*
  moral rights, *118*, *120*
  obligation to protect, *109*
  obligation to report ideas, *117*
  ownership, *109*
  photos and videos, *170*
  pre-existing and new, *113*, *117*, *123*
  risks of social media, *116*
  transfer of rights, *110*, *118*
intellectual property, discuss with
  collaborators, *117*
  guests and visitors, *116*
  students, *116*
  team members, *115*
international training network, *17*, *24*
invention, *124*
inventor, *118*, *127*
invoices, *100*
IP. *See* intellectual property
IPO, Intellectual Property Office, *130*

job application, writing a, *64*,
job contract. *See* agreement, labor
job interview
  Skype, *18*, *25*, *82*
  STARR method, *31*, *73*
job post
  at conference, *13*
  in scientific journal, *14*
  on academic portal, *14*
  on personal website, *14*
  on social media, *14*

job post (cont.)
  to peer network, *13*
  web advertisement, *15*
journalist, connecting and working with, *168*

knowledge transfer officer, *118, 170*

leadership courses, *95*
leadership styles
  autocratic, authoritarian, directive, *57*
  avoiding, withdrawn, distrustful, *57*
  charismatic. inspiring, coaching, *56*
  democratic, participative, compliant, *57*
leave, maternity, paternity or care, *101*
legal officer, *6, 109, 114, 131, 170*
legal processes, mandatory for getting agreement signed, *114*
librarian, *147, 149*
LinkedIn
  connecting to former team members, *74*
  groups, *139*
  job post, *22*
  looking up journalist, *168*
  tool for headhunting, *15*
  your profile, *171, 196*
liquidity planning, *97*
loyalty, lack of, *104*

MailChimp, *193*
managers, how to deal with, *69, 70*
Masters' student, as your assistant, *24*
Matthew effect, *68*
media
  general tips for dealing with, *171*
  preparing for nasty questions, *183*
  preparing your message, *182*
  scandals, *168, 181*
  story telling, *180, 183*
media officer, *169*
media, people involved
  collaborators, *170*
  communication officer, *169*
  funding agency, *170*
  journalist, *168, 173*
  knowing yourself, *170*
  publisher, *169*
  target group, *167, 173*
mediation, *90*
mentor, *60*
  finding one, *55*
  for yourself, *65, 89, 95*
  serving as, *68*
meta-data, *149*
misconduct, *47*
MOOC, Massive Open Online Course, *86, 180*

NC (non-commercial) license option, *142*
ND (no derivatives) license option, *142*
negotiation
  about authorship, *55*
  three steps, *69*
  with top candidate, *35*
notebook, archiving meta-data, *149*

open access, *115*
  articles, *144*
  books, *147*
  green, gold, diamond model, *144*
  sharing of data, *148*
open data, *148*
open research, *151*
ORCID (Open Researcher and Contributor ID), *191*
overhead costs
  calculating, *97*
  not covered by funding agency, *104*

parental leave, *34, 101*
patent, *114*
  application and maintenance costs, *128, 131*
  claims made in a, *125, 127*

how it differs from scientific article, *127*
in which countries and languages, *135*
priority date, *126*, *127*, *130*
protective, *133*
requirements for success, *132*
termination strategy, *131*
what can be patented, *124*
patent lawyer, *6*, *127*
patent officer, *125*
patent processes, mandatory for filing, developing, terminating, *131*
patent, three strict conditions applicable, *127*
innovative, *126*
new, *126*
payments, the timing of, *97*
PCT, Patent Cooperation Treaty, *130*, *135*
performance interview, *84*
personal and professional development, *5*
of team members, *62*
of yourself, *65*
revitalizing yourself, *67*
personality traits
Big Five, *35*
from healthy to unhealthy, *41*
PhD or postdoc
choosing between, *19*
commitment comparison, *20*
costs comparison, *99*
skills comparison, *20*
philosophy, Greek φιλοσοφία, *202*
PIA, privacy impact assessment, of data, *149*
pitfalls
illusion of research excellence, *41*
in choosing name for method, tool, project, *121*
in consultancy, *106*
in dealing with media, *175*, *180*
in involving citizens, *161*
in leading team, *55*
in recruiting team members, *12*

unfinished work from previous job, *93*
plagiarism, *120*
press release
embargo, *169*
tips for writing a, *172*
written by communication officer, *169*
probation period, *87*
procurement rules, *99*, *107*
professor
becoming a, *65*
honorary, *68*
project
account, *100*
budget, *97*
estimating financial value of results, *105*
negotiating free money, *122*
raising additional money, *103*
project controller, *6*, *97*, *100*
project costs
daily subsistence, *107*
direct and indirect, *97*
eligible or not, *106*
legitimate absence of team members, *101*
made before the project starts, *99*, *108*
protecting intellectual property, *107*, *128*, *131*
publication, *107*, *148*
working overtime, *107*
project costs, changing
less favorable exchange rate between currencies, *108*
less favorable internal rates for costs, *108*
moving money between cost categories, *102*
proof of concept, *124*
psychologist, *89*
public engagement, *152*, *173*, *188*
public relations officer, *63*
purchaser, *97*

QR (Quick Response) code, *191*
quotes for goods or services, *99*

222 INDEX

radio interview, tips for a, *172*
recommendation letter
   asking for a, *27, 84*
   writing a, *20, 65, 90*
recruitment policies
   internal candidates may have priority, *83*
   no life-long postdocs, *82*
   PhD graduates should leave, *82*
   recommendation letters may be required, *84*
reproducibility, *49*
   crisis, *41, 52*
   is not enough, *51*
   open data, *149*
research
   types of output, *141*
research assistant program, *17*
ResearcherID, *191*
responsible university, *165*
Retraction Watch, *47, 181*
revenue
   incidental success stories, *128*
   models for dividing between parties, *118*
   university rules for sharing, *103*
role model, *63*
   becoming an academic, *68*
   from within and outside academia, *64*
royalties, *101, 103, 128*

SA (share alike) license option, *142*
sabbatical, *67*
school, Greek σχολή, *202*
scientific integrity
   dilemma game, *53*
   gray zone, *48*
ScopusID, *191*
scouting strategies, *13*
seed money for exploratory work, *99*
selection
   checking for potential bias, *35*
   in case of serious doubt, *25*
   pitfalls, *33*
   STARR method, *38*
selection committee

"'more of me'" bias, *35*
   agreeing on selection criteria, *32, 38*
   completing implicit bias test, *39*
   policies for composition of, *83*
selection criteria, *18*
   expectations of the job, *28*
   fitting into the team, *30*
   funding potential, *29*
   motivation for this job, *28*
   other academic activities, *27*
   papers, talks and more, *26*
   recognition and reputation, *27*
   scientific achievements, *26*
   trainability, *28*
   vision for the future, *29*
self-reflection
   about your own career, *65*
   by job applicant, *32*
   by team members, *40, 59*
   by you as team leader, *55, 72*
   discuss with your team, *51*
SEO, search engine optimization, *199*
servicemark. *See* trademark for products or services
side projects for "free playing", *64*
skills
   development strategy, *86*
   hard and soft, *18, 28, 30*
   leadership, *59, 186*
   transferable, *63, 73*
Skype, *18, 25, 82*
social media
   catalyzing, *152, 177*
   disruptive, *176, 182*
   etiquette, *179*
   search engine, *184*
   using PowerPoint, *178*
   using video, *178, 179*
   working with a moderator, *164*
Socialmention.com, *184*
SPOC, Short Private Online Course (SPOC), *180*
STARR method. *See* job interview
start-up
   budget, *104*
   package, *101*

stipend
  for short stay, *17*, *24*
  for travel, *18*
student conference, *16*
students, policy documents for, *116*
summer school, *17*, *24*
superstar
  becoming an academic, *68*
supervision, limited capacity, *19*
support staff
  being a trusted advisor, *3*
  how to deal with, *70*, *204*

team leader
  being an effective, *57*
team members
  advancing their careers, *62*
  former, *16*, *73*
  keep in contact with former, *91*
  leaving, *149*
team phases
  forming, *58*
  mourning, *59*
  norming, *59*
  performing, *59*
  storming, *59*
tenure track, *65*
  negotiating, *103*
thinking, critical and independent, *43*
  completing assignments only, *43*
  following the mainstream approach, *44*
  risks of running against the mainstream, *46*
  seeing what you expect to see, *46*
timesheets, *100*, *108*
trademark for products or services, *114*, *117*, *118*, *121*

travel
  budget, *101*
  stipend, *103*
TTL, transfer and technology liaison office, *128*
TV appearance, tips for a, *172*
Twitter
  instantly sharing ideas, *116*
  job post, *21*, *22*
  tips for microblogging, *178*

UN, United Nations, sustainable development goals, *166*
URL (Uniform Resource Locator), *190*
USPTO, United States Patent and Trademark Office, *130*

virtual conference technology, *34*

website content
  basic info, *187*
  interests, *189*
  leadership, *188*
  open science, *189*
  outreach, engagement, impact, *188*
  research, *187*
website, personal, *171*
  for life, *185*, *199*
  of researchers, *187*
  of support staff, *196*
Wikipedia
  community effort, *148*
  conflict of interest policy, *201*
  guidelines for academia, *201*
  rules for adding information, *183*
winter school, *24*
work-life balance, *63*